ON THE EDGE OF THE OZARKS

ON THE EDGE OF THE OZARKS

Oral Histories from the Arkansas River Valley

Jason S. Ulsperger, Ph.D.
Kristen Kloss Ulsperger, M.A.
and Kayla Osborne

Foreword by Joshua Lockyer, Ph.D.

To order additional copies of this book, contact:
Xlibris LLC
1-888-795-4274
www.Xlibris.com
Orders@Xlibris.com
141599

CONTENTS

Here is a true story I thought you might get a laugh from . . .

My sister and I worry all the time, afraid we're getting Alzheimer's. She's 82 and I'm 80. She called me the other day and said she had done something so weird, she just knew she had it. I've forgotten what she did. I told her I had done something several months ago that scared me too.

I decided at the time to keep a record of the crazy things I do. So, I bought a little notebook and titled it "Journey to Nowhere." [When I told my sister about it], she asked me what kinds of things I wrote in my journal. I told her, "I don't know. I've forgotten what I wrote in it, and on top of that, I forgot where I put it." That ended my journey right there!

Leona Dorflinger Kloss
Lifelong Arkansas Resident

DEDICATION

We dedicate this book to Ray Kloss. The stories he told, and the way he told them, always had an abundant reflection of Arkansas in them. In addition, we dedicate it to all of the people who work with elders in the Arkansas River Valley. This includes, but is not limited to, individuals employed by local nursing homes, the Area Agency on Aging, Alzheimer's Arkansas, and Arkansas Hospice. Your commitment to the aged certainly deserves more recognition than it normally receives.

ACKNOWLEDGEMENTS

We are highly appreciative of all of the interviewees involved in this project. We know some of you did not grow up in this area, but appreciate your willingness to talk about the path that led you to the Arkansas River Valley. We want to express gratitude to the students who conducted interviews. You all transcribed interviews to the best of your ability. Though we know audio quality of taped recordings and general communication issues cause transcription problems, we appreciate your attempts at definitive accuracy. We also extend thanks to administrators of organizations who allowed students to come into their facilities and carry out interviews. We thank Greg Chappell for allowing us the use of his property for the cover art. Finally, we are tremendously appreciative to anthropology faculty within the Department of Behavioral Sciences at Arkansas Tech University. Joshua Lockyer and Eric Bowne provided a series of questions for interviews relating to the heritage of the region. We believe those questions greatly enhanced the quality of the interviews included in this collection.

FOREWORD

Generalizations rarely represent the complexity of the real world. Brooks Blevins argues that this is especially true of Arkansas. "Outside of Arkansas, writers, social commentators, cartoonists, and filmmakers have gazed upon Arkansaw and have tended toward one of two polarized images . . . : a land of backward and most often buffoonish hillbillies whose very existence underscores the superiority of the Puritan-Progressive strand of American society or the habitat of the upretentious, hearty natural man whose very survival amid the homogenizing influences of a materialistic and hyper-civilized society offers a glimmer of hope for those still harboring alternative visions of life in America, an antithesis to the American myth" (Blevins 2009:193-194). One of the greatest values of oral history collections such *On the Edge of the Ozarks* is that they belie the falsity of stereotypes in favor of the details of remembered lives.

Arkansas might benefit from placing greater value on our elders' remembrances of their lives. In other cultures, past and present, elders are often revered as repositories of wisdom, their life histories and experiences the embodiment of valued lessons that can help succeeding generations navigate life. In recent decades, characterized by shorter attention spans, faster technology, and ever more absurd gimmicks for capturing young people's attention and directing their energies, we seem to have cast history itself into the dustbin of history. This third volume of oral histories generated by the Ulspergers and their students as part of a nascent interdisciplinary program in Ozark-Ouachita Studies at Arkansas Tech University is a welcome counter to this trend. My hope is that collections like this gain widespread use and increasing traction as the people of the Ozark-Ouachita region work together to better understand their own history and build an enlightened, empowered, just, and sustainable future upon the memories contained in it.

Joshua Lockyer, Ph.D.

13

PREFACE

We All Grow Old

I usually begin the semester in my social gerontology class by stating, "We all grow old." People who study aging realize such a phrase is multifaceted. Aging is something that affects our biological characteristics, psychological wellbeing, and social relations (Quadagno 2014). *On the Edge of the Ozarks* is a collection of oral histories from elders living in the Arkansas River Valley that touches on each of these areas. Like many people in our culture who are aging, elders in the River Valley miss their youthful skin, have problems remembering things, and are struggling with loved ones who need long-term care. However, they also realize the pleasures of having a life full of cherished memories, the positive sense of self that comes from being a grandparent, and an abundance of leisure time.

I started teaching social gerontology when I took a sociology position at Arkansas Tech University in the fall of 2006. I never taught on the topic before. However, I did have a personal, academic, and research background in aging related issues. As a teenager, I loved to visit my great-grandmother, Effie Blasingame. Though I remember visiting her at the various houses she lived in through the years, most of our visits took place in nursing homes. Before dementia took away her vibrancy, I loved seeing the snuff sneak out of the side of her mouth, while she recited short songs from her childhood. I can see myself chuckling at the corner of her bed while she sang one of my favorites, "Two little boys walking up a hill. Their butts cracked open like a sausage mill." Those visits created a soft place in my heart for the elderly. They also led me to intern at a Jonesboro nursing home while I was working on a graduate degree in sociology at Arkansas State University.

I thought I wanted to be a nursing home administrator, but soon realized that the business side of working with the aged, emphasized by the director I was working for, was less appealing to me than the social psychological side. I loved visiting with residents. I could not wait to get into their rooms, hear about what it was like for them growing up, and indulge with them on fond memories. Maybe it was because my wife and I were in a new town without many acquaintances. Maybe it was because I longed to talk to my great-grandmother who passed a few years earlier. Whatever the reason, I think the experience was good for me, and I like to think it benefited the people I visited with as well. When I entered Oklahoma State University's doctorate program in sociology the next year, there was little debate in my mind as to what my dissertation and subsequent research would focus on—nursing home residents (see Ulsperger and Knottnerus 2011).

We all grow old. I do not think that many of my social gerontology students give that idea much thought before I say it the first day of class. We work through the semester discussing changes in cellular structure, altered thinking ability, and retirement related issues, but for a certain percentage of students, aging is still something that does not seem real. To battle that, I have them go out and interview someone from the River Valley who is 65 or older. When I was younger, I do not think growing old was something I thought about much, until I visited with my great-grandmother. Some young adults do not depend on, and subsequently do not visit with, elders as much as in the past (Ulsperger 2001). Even if they do, they do not ask deep questions about the lives of elders. Interviewing someone over the age of 65 gives them intergenerational contact and a better connection to what we talk about in the classroom.

I decided quickly that taking the interviews and publishing them in a collection was a great idea. It would give elders a documented source of information to pass on to family members. Moreover, donating the collections to the Pope County Library would preserve stories from the community that would have otherwise slid through the cracks of time. The collection of interviews from students led to two previous oral history projects I helped to edit—*Voices of Pope County* (2008) and *River Valley Reflections* (2009).

Students and interviewees benefit from an oral history collection like *On the Edge of the Ozarks*, but this book is for anyone who is interested in the elderly, historical information, or Arkansas culture and its people. I also encourage the casual reader to note the structure of each interview. I hope you are inspired to collect your own oral histories from family or community members. As reflected in this book, you do not have to be overly technical to get some good information. If you decided to collect your own oral histories, I recommend you check out a piece Kristen and I wrote for the Arkansas Library Association's journal (see Ulsperger and Ulsperger 2011). It offers a simplistic, systematic guide for the oral history novice.

I want to close with some information I pass along in every one of these collections. First, the editors finalized all of the interviews in this book to reflect the actual words of interviewees to the best of our ability. From interviewees mouths, to interviewers recorders, and then to type, we realize inaccuracies emerge. We apologize to participants and their families for any mistakes. Interviewers were instructed to transcribe statements to reflect interviewee's dialect. Sometimes they did. Sometimes they did not. Therefore, inconsistencies with certain words throughout the book exist. Second, an oral history does not encompass a person's total life. The chapters in this book just represent pieces of each person's existence. Regardless of these issues, my editors and I hope you enjoy acquainting yourself with each interviewee's life as much as we did.

Jason S. Ulsperger, Ph.D.

CHAPTER 1

Take Away the Ipods

Biographical Information

Name: Alan Lee Tompkins
Place of Birth: London, England
Date of Birth: December 11, 1944
Interviewed By: Toni Cook

Can you tell me about your birth?

My father was an American pilot in the Air Force stationed in England in World War II. My mother was an English war bride. I suppose my parents got carried away one night while they were still dating. It was only a short time before my mother realized I was on my way. They were soon married and were married for almost 54 years until my father's death.

What was your childhood like?

My childhood was special. I grew up on a farm in Northwest Missouri. I went to three different one-room schools. When I was 12, we moved to Kansas to a fairly large city with a population of 20,000. I started school there in the 7th grade when I was 12.

Had you been in school before that?

I started school in Missouri when I was four. [I started so young] because we lived next door to a one-room school that I played at every day. The teacher, Miss Davis, who was about 17 at the time, told my mother that since I was there all the time anyway, I might as well be in school. She was right.

What is your favorite childhood memory?

I have so many wonderful memories as a child. It would be hard to place a finger on one incident in my life . . . My fondest memories are centered on the good times I had with my dad, who is now deceased. I grew up in an era of hotrods and fast cars. My dad was a mechanic, so I got to spend lots of time looking at and working on the muscle cars of that time. He died from cancer several years ago. When I look back on my childhood, the times he spent with me were special. He was a special person in his own way. He was patient, kind, firm, and very protective.

What was a typical day like for you when you were growing up?

I will pick a day from 1949, when I was five. The day would start by having my dad get me up at sun-up to feed the hogs, gather the eggs, and feed the chickens. I would eat breakfast and walk to the one-room school across the road. After school, I would walk home and do my chores again. Only this time, I had to walk to the back of the pastures to get the stubborn cows to be milked. I would gather the eggs again and help my mom run the cream separator from the milk my dad brought in from milking. Then, we would eat supper and dad would help me with my schoolwork. I did my homework by oil lamp until I was 12, because [by then we had] electricity.

What memories do you have of your parents?

Memories of my parents are fondly remembered. I got along very well with both my mom and dad. I went through the stage where "your parents are dumber than stumps," but quickly realized as I got older, that they were actually very intelligent. I found out that parents are great to have around when you need a shoulder to cry on or someone to help you. My parents were always there for me. As I get older, I hope my own children can think of me in the same ways.

Will you discuss your education?

My education was a series of interesting events. Early on, as I said, I attended school in a one-room blockhouse, about a stone's throw from our farmhouse. I told you the teacher was a young girl I remembered as Miss Davis. My third year was in a schoolhouse about three miles from my grandmother's house. We had moved to live and farm at my grandmother's farm. After that, I went to a school close to a small farm we moved to when I was 10. My dad tried farming two places— my grandmother's and our farm we were renting from a Mrs. Wilson. I had to walk to school a lot when the weather was good. In the summer of 1956, we moved to Great Bend, Kansas, where my dad took a job working with his brother-in-law. Farming had become so bad; we had to move to the city to try to make a living. There, I attended the Jr. High and then the Senior High. I graduated with 12 years of regular education. I never attended college.

Where have you worked?

I have worked in many places and different fields of employment. I have worked for the State of Kansas for the Transportation Department, and The United States Air Force, stationed in places like Arkansas, Texas, Indiana, the Philippines, and Vietnam. I just retired from a French company called Dassault Falcon Jet. I worked as manager over the paint department for nearly 30 years.

Tell me about your spouse.

At first, she was a cute girl who I thought was fun to be with and she had an upbringing very similar to mine. As the months went by, we dated several times and I began to realize that she was the person I wanted to be around forever. Like most couples, we talked about our expectations towards our future—what we expected from life. We came to the agreement that our goals would be best met by joining forces. We got married to fulfill our ambitions. We have been married for almost 45 years. We were married March 26, 1966, in her older sister's home, with just a few friends attending and of course some relatives. I wore my Air Force uniform, and she wore a beautiful white dress outfit. It was a small wedding—no hoopla or band, no buffet or anything, [just] the traditional cake and punch. It was small but it was perfect. It was one of the best days of my life.

Tell me about your kids.

We were blessed with three beautiful children, two boys and a girl. We have Patricia Joan, 40, Jeffery Alan, 39, and Stephen Ray, 37. They [are all] wonderful people. Patricia Joan works in the accounting department of Dassault Falcon Jet. Jeffery Alan is currently in Special Forces and in his second deployment to Afghanistan. Lastly, Stephen Ray is working for the railroad in Houston, Texas.

Do you have grandchildren?

Patricia Joan gave us two beautiful grandchildren—two girls, Katie Rose Elmore and Abby Leigh Elmore. Katie Rose is a spitting image of her mother. She has a temperament that is like her mother. She is the cutest thing you will ever see. Abby Leigh is completely opposite. She is a force to be reckoned with. She looks like her dad somewhat, and in her eyes, her mom. She wears clothes that most kids do not wear for a couple of years down the line. We are hoping that our sons will eventually marry and have children someday. It would be nice to have a grandson.

What was it like to have three children?

Having three children was more of a burden on my wife Mickey than me. I worked all the time, and Mickey took care of the kids mostly by herself. I helped late in the evenings and on weekends. Three kids, especially as close in age as they are were tough, very tough. All the kids were very active. Jeff and Steve loved to scare their mother and their sister every chance they got. When the kids were smaller, Mick had three to rock to sleep at the same time. Feedings and diapers were a challenge . . . Most of the credit [for raising the kids] should go to my wife.

What brought you to Arkansas?

I came to Arkansas in July of 1965. The Air Force stationed me at Little Rock Air Force Base in Jacksonville. As I previously stated, it was then when I met my wife, who is from Arkansas. I could not think of ever taking her away from her family.

Can you discuss your years in the service?

I am a veteran of the Vietnam War. I served four years in the Air Force from January 1965 until February of 1969. I was in Vietnam for two different tours and spent 16 months in the Philippines. I was a mechanic working on airplanes.

As you have gotten older, what are your favorite things to do?

Some of my most favorite things to do are just to be able to relax around the house. I like going to [yard] sales with my wife and tinkering around the house . . . I enjoy that I don't have to hurry to get things done. I can spend my time on things. Spending time with my wife is a favorite pastime. We like to travel and goof off. Some things never change with age . . . My favorite hobbies are spending time with the grandkids, swimming, and skiing. All three are fun for me in different ways. I have loved swimming and skiing since I was a young adult . . . There is nothing better than getting to spend a day with my grandkids at the park or teaching them how to do something. Nothing is more rewarding.

What do you think about aging?

The best thing about aging is the fact people do not expect that much from you anymore. It eases the stress that a person has carried for most of their life. When it comes to aging, some of the downsides is the soreness you have you didn't have before, the wrinkles, my hair color keeps changing, loss of eyesight, hearing, and all your friends seem to forget your phone number. [Overall, I think] my aging experience has been one of a mostly positive nature. I have seen so many positive things in my lifetime. I have had many good experiences. I learned [the value of] hard work. [I have seen] the invention of medicines that have been developed to cure or eliminate sicknesses like Polio. [I remember] the walk on the moon by Neil Armstrong . . . It was the height of the American power.

How is the world different now than it was when you were younger?

The world is so much different now than it was years ago . . . I remember getting on airlines without as much as a screening device. The crimes are so much more violent now. You cannot travel to places overseas and feel safe. Travel is a hassle, politics have become a separation of people, and it just isn't anything like before. War was something fought on other grounds . . . Life changes and so do people. [The younger generation now] is lost. You can't get correct change from a kid behind the counter at McDonald's if the computerized cash register is broken. If you took away the IPods, cell phones, and texting, the generation of today could handle adulthood. The younger generation could survive if we trained them better. In my opinion, the younger generation of today will have to answer for the screw-ups the "middle generation" is doing today.

Would you change anything about your life?

Looking back on my life, the one thing for sure I would do differently is go to college and get all the education I possibly could. [I would have] traveled more to see places not possible today due to dangers and wars. I have had a good life and do not regret my past, I may want to change a few things, but I am happy with what I did do.

CHAPTER 2

Tall, Red Headed, and Nice Looking

Biographical Information

Name: Barbara Ann Rae
Place of Birth: St. Joseph, Missouri
Date of Birth: October 14, 1948
Interviewed By: Jason Amaral and Bobby Davis

So, how long have you lived in the Arkansas River Valley?

Since 1962 . . . I grew up with what I consider a perfect childhood . . . My mother didn't work, so she was home for me all the time. I had a tree house and a swing set . . . no broken family. When I was born my mother said I was goin' to be named Barbara Ann, but I was goin' to be called Bobby—all my growing up years I was called Bobby or Bobby Ann. In school I was Barbara, everywhere else, I was Bobby, and I liked it.

Tell us more about your mom, some about your father, and something about grandparents.

My mother was very smart, dark hair and dark eyes, very religious, very sweet, and supportive. She didn't care about *things*. She didn't care if her dress matched her shoes—she didn't care about *things* at all. She was always there for me, and anyone that needed her. I was always a daddy's girl. [He was] very gentle. I never met anybody that didn't like my dad. My dad had a sandy colored truck. We would go on drives a lot. My grandmother's name was Patty. She's where I got my spiritual upbringing. She's one of the finest people I ever knew. I think I was her favorite. She volunteered at the [Arkansas State Tuberculosis Sanitarium]. I really learned a lot from her. I miss her. We were very close. [I was in] college when she passed. She had a sister named Molly, and they both married twin brothers. She would take me to the library. She's the one that introduced me to the library . . . Oh, we had such a good time [when we would] read books, play solitaire, and put puzzles together. [My grandfather] liked to play cards, and he was the town pool shark. He would go to the pool hall and play all the time. I wasn't as close to him though. I had another grandfather though. [He] was so much fun. He wanted me to be named Carmen Miranda after the actress. He taught me how to play the harmonica. He did magic tricks and took quarters out of my ears. I can tell where his sense of humor went on to my brother and my son.

Do you have any brothers or sisters?

I have one brother; he is four years younger than I am. I remember he was always taller and bigger than me and everyone thought he was the oldest one because of our size differences.

Do you have any kids?

I have a daughter, a son, and a stepdaughter. They grew up on a farm out in the country. I think there were a lot of advantages to that, disadvantages too when they got into a bigger town situation. My son still lives on the same farm, and my daughter has her master's and she works as a counselor. I have a grandson too. His mother is from Russia, so he's already speaking two languages. He's just the light of my life. I bought him a coat, hat, boots that all say fire chief on them and a little umbrella to match. He was really proud of his little outfit. I don't know what he's going to be for Halloween, but I can't wait to see the pictures. He's got a baby sister on the way, so that's going to be fun. I have another grandson that's a sophomore in college this year, and I'm very proud of him too. I have a little granddaughter that's about five months old now that I'm hoping I get to spend more time with.

What kind of chores did you have while growing up?

Like I said, my mom stayed at home, but she taught me to cook. I feel I have always known how to cook by being in the kitchen with her. My mom cooked dinner every night, and I would help her cook. She made the best chili. I wish I could make chili like she did. I've never made a pot of chili that tastes like hers. She was one of these that didn't care if she cooked from scratch. She could make piecrust from a mix, pudding from a mix, cake from a mix. It was always good but it wasn't always from scratch. I always had to do the dishes. My brother got to empty out the trashcan and start fires in the trash can. If I didn't do the dishes, they stayed there until I got to them, but if my brother didn't take out the trash my mother would do it for him . . . She wasn't fussy about how I kept my bedroom. She was one of these who if company was coming then we needed to clean our bedrooms and get the vacuum going. She wanted us to look presentable when we needed to be.

Talk about food practices unique to the area you are familiar with from living here.

My grandparents had a garden, but I'm not sure how they got their seeds. I enjoyed the food [they grew]. We hunted mushrooms.

How was school for you?

I walked to school every day. It was a pretty good ways—about ten blocks. My mother walked with me until my brother started walking to school with me. Then, I would come home and read then have dinner and went to bed. On the weekends, we would go visit my grandparents, always on Sunday. My family was very close. My cousins were like my brothers and sisters. I started the 9th grade at St. Joseph High School in Missouri, and then I moved to Van Buren, Arkansas. It was like starting life over again. I didn't hang on to any old friendships. I was excited about the new school and the new friends and the new house and the new job for my dad, but Van Buren's school was much smaller than the one I went to in Missouri. I wore glasses, but I was never picked on or teased about it. [I went to college] at UCA in Conway. It was Arkansas State Teachers College when I started, and then it was State College of

Arkansas after that. I started out in Journalism and English [and then changed] my major to Elementary Education and Teaching.

Do you have any teachers that stand out in your mind?

Emma Lee Lewis was my teacher [for three years]. She was very strict, and she was very fair. She and I became friends when I wanted to become a teacher. By then she was ready to talk about retirement, because she was ready to retire, but we stayed near each other always. She gave me a lot of tips when I started teaching. I had one teacher for English that accused me of cheating and made me sit by myself for an entire year, but I don't think it was because he didn't like me, he just didn't trust me . . . I had a teacher for journalism that was really good about supporting me.

Can you talk about movies and television shows?

The first movie [I saw] that sticks out in my mind is *13 Ghosts* (1960) . . . You had to watch it in 3D. I think they've remade it now. I like *Fried Green Tomatoes* and *Steel Magnolias*. I like the relationships movies. I love *Survivor*. That's my very favorite show. I won't miss it. I don't know what it is about it, but I really enjoy watching *Survivor*. I don't think I've missed two episodes in all the seasons. I really like *Survivor*, and I really like *The Big Bang Theory*.

Do you have any thoughts on romance?

[People meet on the Internet] and I've seen it work out. I've seen it be a disaster for people too. It wouldn't be something for me to do. [I remember my first kiss] was with Jerry Lasper—tall, red headed, nice looking kid. I think I was a sophomore in high school.

When you were growing up, what was your favorite holiday?

Christmas. I like to shop for people. I love to give presents. I love the Christmas music and songs, the going to church, the religious Christmas music. I love everything about Christmas. I love the snow and the winter season. One Christmas my cousin Tommy got a BB gun, he was so proud of [it]. I could shoot it better than he could. Being younger and a girl, it took some of the fun out of getting a BB gun for Christmas.

Do you remember your first job, and did being a female have any impact on your employment?

My first job was working for the Girl Scouts. I worked for the Girl Scout office filing and doing office work. Then, I would work at the camps for the kids. I spent one entire summer living in a tent. It didn't pay much, but I loved it. For a while, I thought about becoming a professional Girl Scout until I did my practice teaching. I think the majority of teachers were women. I don't know if that's changed over the years, but that's how it was when I taught. I never had any trouble . . . I have no problems with that. [In terms of hard work], I see people come put in the hours and cry about the job. I don't think it's like it was when I grew up. I didn't know anyone when I grew up that received anything from the government . . . Everybody talked about their jobs and was proud of their jobs. I see a lot of that now has changed.

How would you feel about the government increasing the retirement age?

I think it's a personal decision for when you retire. For me, I don't think they should raise it, because I'm getting close to that age. I can't really see any benefits in raising it, except that some people could work longer. I think there are a lot of older people who have to work. For that reason, they should be allowed to work for as long as they're able.

What is your perspective on spirituality?

I don't go to church, I do read my Bible. I have very strong faith. I've seen things happen in my life that I know came from God. They happened because of God. So, I would say I don't think you have to be in organized religion to have a strong faith. I grew up going to church and Bible school, so I have that background. [As I have gotten older], my faith is stronger. I think that you don't really think about it on a daily basis when you're growing up. The older you get, the more you see things that bring you closer to God.

Now how do you feel about being an Arkansan?

I've been everywhere in the United States except for Maine and Washington. I like to travel. I like to see all of the things that make the United States what it is. I love the caves and everything, but I love Arkansas. I think there are lots of advantages to being in Arkansas as far as the nature and things you can get out and do. I think other states don't have as much respect for people from Arkansas as they should. As far as sports and other common interests, Arkansas people have a lot of interests that they have in common with each other. I think when I first moved [to Arkansas stereotypes] were truer. There were more people that stayed in the hills and didn't come to town very much. It's a different Arkansas now than when I moved here.

Speaking of change, attitudes about race have changed in recent decades.

I think that's a very good thing. I think you're a product of how you're raised, and I think people are raised now to be more accepting of everyone. It wasn't that way when I grew up. There was a line between black and white. The town I grew up in had their own black school. The town my grandparents lived in had no black people who lived there. Things are not separate now like they were then, and everyone seems to work together. I think it's good. I think in the past if you didn't go to college then you weren't exposed to things, the art, the music. I think things are more accessible now than they were. So, I think society is more educated than they were about things other than just every day routine. I have a lot of hope for society. I think people are more accepting and rounded in their interests. We have a wider field to pick from . . . If you didn't read books then there was a lot of things you missed, but people don't know their neighbors . . . I think that there's a closeness there that we are missing out on. I think families are maybe as strong as they were back then, but I don't think neighbors are. I think friendships now come from the workplace instead of the neighborhood. I don't know if that's a good or a bad thing.

What is your opinion on growing older?

Now is my favorite time . . . You don't have a choice, you have to grow older, but you don't have to grow old in the way you think and the way you act. Growing older limits you in some ways, but it's also free because you don't have the responsibilities of family and children. You're financially maybe to a place where you don't have to worry. There are a lot of good things about growing old until your health starts to fail.

Then it's a whole other story. Being limited by your health is the hardest part, [but overall] the experience that you have, that is the best part of growing old.

Do you have any advice for future generations?

I would like to see stronger marriages. I would like to see people make a bigger effort to keep it together. Sometimes it's impossible and you can't, but I would like to see people not be selfish for the sake of their children.

CHAPTER 3

This Is Where Jimmie Lou Is

Biographical Information

Name: Bobby Joe Trusty
Place of Birth: Shiloh, Arkansas
Date of Birth: October 12, 1932
Interviewed By: Holly Woidke

Where did you grow up?

[I grew up in] Shiloh, Arkansas, in Pope County. We moved to where I grew up when I was about four, but we lived within a couple of miles of where I grew up all my life. I was actually born on Highway 7 just past the VFW club out there, on the left, past there. I have a twin brother and we always played together, always had a lot of fun, I guess. We did whatever. We had tricycles, later on, of course, bicycles. One thing that I remember about playing was that we, I don't know why we did that, but we was always [pretending to] have a store, you know? We'd have a store, and we'd use big green leaves for money. I remember doing that, playing like that. My youngest brother is several years younger than me. I don't really remember playing with him that much, because we was older. Of course, my sister, she never did play with us much. I guess maybe me and Bill just played with each other so you know, that was what I remember. Of course, we lived on a dirt road, but that didn't hinder our bicycle riding. We rode our bicycles all the time. Grandma just lived close to where we lived, and we'd cut through the field and go over to her house all the time . . . Then, my uncle lived just right down the hill from us for years, and they just had one boy and he was my sister's age. I never do remember playing with him any, but anyway he lived right down the hill from us. It was a good way to grow up, I guess.

How many generations of your family have lived in the Arkansas River Valley?

Well, my great-grandpa came here from Tennessee . . . My grandpa and his wife had lived in this area for all their married life. 'Course that don't seem like that many years ago, but that's been years ago. So, we've been here for I imagine, close to a hundred years, maybe even longer. In fact, my daddy was born in the 1800s . . . The Trustys lived in Paris. That is where my daddy come from. 'Course, there's a lot of Trusty's that still live around Paris. None of them are close kin to me. Daddy only had two brothers, and one of them died pretty early in life [and the other] stayed in Paris.

What kinds of foods do you remember your family growing for themselves?

[We grew] potatoes and other things.

Did you save seeds or trade them with your neighbors?

I don't think [we] ever traded them. 'Course [we] saved seed—tomato seed sometimes if we had outstanding tomatoes. 'Course I know we used to save the cantaloupe seed and the watermelon seed, maybe squash . . . The seed back then didn't cost hardly anything.

Was your family largely self-sufficient in food production?

Yeah. One reason I say that is that for years we killed hogs every year. We'd usually kill four hogs, you know? Cut it up, and I can remember us smoking it. We did that for years and years. You wouldn't kill the hogs until it got cold. A lot of times we'd kill them on Thanksgiving. You can't imagine how that is, having fresh lard that you'd made yourself. I can remember Mama cutting up the lard, cutting up the fat pieces, and then we'd start a fire around the wash pot and render that lard in the wash pot. I remember getting a piece of tenderloin and just putting it in there while she was rendering the lard, and that would cook it like that. 'Course you had to get it out before you got through rendering the lard but we'd render the lard and pour it into five gallon buckets. I would say that we was pretty well self-sufficient there for several years.

Have any of your children or grandchildren carried on these food traditions?

Of course, Cindy does. Cindy's the only one that still makes a garden.

Going back to your childhood, what do you remember about your mother and father?

Of course, I remember she always cooked, you know? Always had a meal, breakfast and dinner—dinner being the noon meal, and then supper . . . She always cooked and we'd always enjoy the meals. In the summertime especially, we'd have fresh vegetables out of the garden and tomato soup and just whatever. I remember that. I never do remember her driving when I was small, but I guess surely she did . . . Later on, I remember her driving. She had a car, and she'd go wherever she wanted to go. She never did work . . . Of course, I remember Daddy working all the time. I remember him working at Plunkett—Jarrell's. That was a wholesale grocery company and he worked there for, I don't know, several years . . . Of course, I remember him talking about farming when they first got married, but then he eventually left and worked at that wholesale grocery.

What did that involve?

He delivered groceries—whatever the people bought. Of course, the thing I remember about it, when me and Bill got a little older, you know, like maybe 7 or 8 or 10, I don't know, he'd let us go with him. He'd go up Highway 7, go up into the mountains, and we loved to go up into the mountains with him. Of course, we'd help him deliver stuff too. You know, stuff that we could carry. I remember we never did feel like it was work. We felt like we was having fun . . . [Now] it's been paved for years, but back then it wasn't paved . . . It wasn't paved until after the second war was over, so you know, the late 40s sometime. Maybe early 50s, I don't know.

Did your father always deliver groceries after leaving farming?

[He] felt like he could lift anything, and he got a hernia from lifting something that was too heavy for him . . . The doctor told him that he'd have to quit. So, anyway, he didn't hardly know what he was goin' to do, and he was acquainted with Elmer Ethridge. Elmer had a thing kind-of like groceries except that it wasn't any heavy stuff. Daddy used to say it was "razor blades and aspirin tablets." Anyway, he told Daddy that he'd give him a job selling, and of course Daddy was real, he was real friendly, you know? He got along with everybody and was really a talker, so he was ideal for a salesman. I remember him talkin' about they owed him $75 a week. They'd pay him $75 a week, or they'd pay him commission, whichever one he wanted. He said, "Well I'll take $75 a week" and so after he'd worked a couple of weeks, they said, "Well, we owe you a hundred and fifty dollars for two weeks, or we owe you two hundred if you work on commission." He said, "I'll just work on commission." So, after that he always worked on commission, you know, he was really, really successful at that, you know? We was primarily going on the same routes that he delivered groceries two years before, and so a lot of those people he'd known before for years, you know? He worked for that Elmer Ethridge for several years, but then [Elmer] started trying to change everything . . . Old people don't like change, so him and this other guy started the Joe Paulk Wholesale Company [which distributes groceries and other commodities].

How did that work out?

Daddy was very conservative with his money. He always saved his money. Of course, he never did make no money to amount to anything until he started selling. Anyway, Joe offered him partnership, and it was just going to cost like fifteen hundred dollars, which Daddy had the money, but he told him no. He didn't want a partnership, but eventually the company got really big and his fifteen hundred would have been a really, really good investment, but he didn't ever do it . . . My twin brother was living in California, and he moved back here and worked for Daddy for a year. He took [Daddy's job after he retired]. Of course, it was almost like they didn't realize Daddy had even quit . . . Anyway, Daddy always smoked and he had emphysema pretty bad I guess . . . Later on, he got to where he would just sit at the table, and that was about all he did and when you went in to see him, why, he never would say anything. But, if you asked him something, he'd answer you but he just wasn't interested in talking . . . He died when he was 76. They said his heart . . . He struggled so hard to breathe that it strained his heart.

What do you remember about the people you came across on Highway 7?

Daddy had a nickname for everybody. Hankins [was] Tom Hankins. He had a grocery store at Pelsor, and he was a really colorful individual. Him and Daddy always kidded and talked to each other . . . One time Daddy hit a hog of some kind, not a big one, but a pretty good size one. It was down the road, not too far from Hankins's store, so when he got up there to the store he told Tom about it, and he wondered if he knew who it might belong to. [Hankins] said, "Nah, I don't know who it belonged to. That's just an old mountain hog, don't worry about it. Just forget about it."

Does anyone else stand out?

There was a woman running a store up there, and Daddy called her the "Wild Woman." I don't know why he called her that, but that's what he called her.

Can you talk more about your siblings?

My oldest brother was eight years older than me. My next sister was about less than two years younger than him, so they had two right close together . . . Then, my next sister, she's 82, she's about almost three years older than me, and Bill, that's my twin brother . . . My youngest brother, he's 74. [When] I graduated from high school in 1950, that's when my youngest sister [Linda] was born.

What did you do after graduation?

Of course, I joined the service. I graduated from high school on a Friday, and I was sworn in the Air Force on Monday . . . When I came back through town, they took me to Fort Smith to be sworn in, and then we came back through town to get to Little Rock . . . [Then I] got on a train to San Antonio. That's where I went for basic training. After basic training, I stayed in San Antonio as a "permanent party." That's what they called it after you got out of basic, and I was in the motor pool, drove staff cars and everything.

When you were outside of Arkansas, what did people let on as their perceptions of the state?

I don't know. I think they mostly had the perceptions that we was barefooted and pregnant. Somebody was always saying something like, "You mean you wear shoes?" I've always liked Arkansas. Of course, one reason I think is because of the scenery, the trees. 'Course you don't really appreciate it 'til you're gone.

Backtracking a little, where did you go to school?

You know where the VFW club is? That's where I went to school, Baker's Creek school. [It] was a grade school. I went there for about two and a half years, and then [the original building] burned . . . I guess it was completely destroyed, but anyway, they built it back. In the meantime, we went to [Arkansas] Tech. They had a grade school there at Tech, and I went to school there while we was out of that building. Then, I went back to Baker's Creek, but I only went one year then because I had to come to town for the 5th grade. I went to what was called North Ward, now it's Crawford School, and Ms. Crawford was the principal of the school when I was in the 5th grade. I went from there to junior high. Junior high was on Arkansas Avenue. Do you know what that school looks like over there now?

The one that is boarded up?

Yeah. Well, see, then the seventh, eighth and ninth was on one end of the building, and the tenth, eleventh and twelfth was on the other end . . . Then the auditorium was in the middle, so that's the way it was. I don't remember how many people we had that graduated . . . Me [and my wife Jimmie Lou] got to walk together . . . We was dating then.

Did you have a car?

We had cars. We had one at first, but it wasn't for me and Bill to use. We finally talked Daddy into letting us, and he said, "Well, I'll let y'all use the car once a month a piece." We could have a date or do whatever. We tried to get him to let us have it twice a month and he said, "Well, no, I couldn't do that. That would mean the car would be gone every week!" We [eventually] had two cars. Back then, back in the early 50s, well it was '49 actually, they wasn't making very many [cars] yet, but Daddy signed up for a Ford and a Chevrolet both. He got the Ford first, and then the Chevrolet place called him and so he bought the Chevrolet, too. So, my older brother took the Ford, but anyway, we had two cars after a while. I drove the Chevrolet, and it was brand new and I drove it out to Jimmie Lou's place to pick her up [for the prom]. That was off of 124, and 124 wasn't paved. It was raining and boy, that car was a mess. I remember. That was in 1950, so I remember that. [We met] in junior high . . . I worked in the cafeteria when I was in the 9th grade. Her mama worked in the cafeteria. 'Cause her mama didn't know how to drive, Jimmie Lou would drive the car to school at 14!

Can you talk about your wedding?

I went overseas in '52. Back then, you didn't come home. Whatever your tour was, you stayed . . . I went to Japan. I was stationed there for eight months. Backing up a little bit, Mama and Daddy had to sign for me to join the service, and the Korean war started about three weeks after I joined . . . Anyways, we went overseas to Japan in '52. We was over there for eight months. You know, that seems like a long time, but I was reading something the other day that they signed the peace treaty with Germany really late, like in the '50s . . . They didn't sign the peace treaty with Japan until '52 or '53. After they signed it, they shipped a lot of us out of Japan, so they said, "You want to go to Korea or you want to go to Guam?" and I said, "I'm not in no hurry, just send me to Guam." I stayed seventeen months on Guam, so you know; I was gone for a little over two years. After we got back, I just had six months to do, and I was stationed at Langley Field in Virginia . . . That's where the CIA is, and of course, Norfolk was just right across the bay, and Virginia Beach. Virginia Beach is now a great big place. When I got discharged there in '54, it was a beach. That's all it was. Wasn't no town . . . Anyway, I came on home then, and me and Jimmie Lou started talking about getting married [since her first husband was killed in Korea]. She said, "Well, we'll get married. We'll get married on the 15th of October." My birthday is the 12th of October, and she was already 22. She didn't want to see her name in the paper with her 22 and me just 21, so she waited until I'd be 22 . . . Back then, most people, if they'd been married before, just got married. They didn't have a really big wedding . . . I don't remember how many people was there—maybe 10. Jimmie Lou's sister was there, and her best friend was there. Wasn't nobody there for me that I know of, but I didn't care. Neither one of our parents was there. We [were] married at 6:00. When we was leaving town, we was driving across the Dardanelle Bridge and I looked across at my watch . . . It was 6:25. [We have] been together 58 years so far. I guess we won't get a divorce. It's too late now!

Tell me about your children.

Well, of course, Cindy's the oldest and Tammy is next. Cindy was born in 1957. I was going to [Arkansas] Tech then. I graduated in '58 and she was born in January of '57. Of course, I graduated in '58, so it wasn't very long. Seems like we had bought us a house before she was born—over on H Street. That was just a nice neighborhood then. When Cindy was born, the GI Bill started payin' me $160 a month, and Jimmie Lou was working. I had a job keeping books for a place here in town. They paid me $25 a month. I'd just go down there maybe twice a month and do their books, but our house payment was $33.30 a month. So, $25 almost paid the house payment, but anyway, seemed like we had as much money then as we ever had . . . Anyway, Cindy's 55, Tammy's 52 and Jim Bob is 47. When Jim Bob was born, he was born early in the morning. The nurse comes walking out with him and said, "Well, you have a red headed boy." I said, "Well, I don't care what color his hair is. It's a boy!"

What about your grandkids?

[We have] seven grandkids—two girls and five boys. [The youngest is] 15 and the oldest is 32.

What is your perspective on spirituality?

I had always went to church. I got saved when I was young—baptized and everything . . . I was baptized by immersion [dipped in water as opposed to sprinkled with it]. My uncle was a preacher and a Presbyterian. I went to Cumberland Presbyterian Church. My uncle baptized me . . . When I was in the service, I went to chapel some, not that much. After we got married, Mama's preacher tried to talk me into coming back to Cumberland Presbyterian. I didn't really have any ties there other than my mama went there. Jimmie Lou was real active in her church [First Free Will Baptist], so I just started going there. I rededicated my life in November of the year when we got married. I've been active ever since. I started teaching the college [Sunday school] class. Then, as I got older, I started teaching young married people, then the older married people. I taught Sunday school for [around] 45 years or something . . . In the early '60s, I was ordained a deacon.

As you've gotten older, have you had any health problems?

I had a torn aorta, and people die from that all the time. I knew there was something wrong with me. I was out at Wal-Mart when it hit me. I thought I was goin' to pass out, but I didn't. I got over it enough that I went and got gas because that's why I was out there. I got home and waited for Jimmie Lou to come home, and so she come home and I told her to take me to the doctor. He did an EKG and said, "Well, you're not having a heart attack." He said, "I don't know what's wrong with you. You want to go to this hospital or do you want to go to the Heart Hospital?" I said, "I want to go to the Heart Hospital." He said he'd get an ambulance, but I said, "No, Jimmie Lou would drive me." I had a stent already and the doctor said, "Well, I've got good news and bad news. Good news is your stent is fine but you've got a torn aorta." He said, "We're goin' to have to operate." I said, "When are you goin' to operate?" He said, "Immediately." I didn't know nothing for three or four days. It was quite an ordeal. I told Jimmie Lou one morning there that everything was shutting down and they said, "We'll have to put him on dialysis if he doesn't change." My kidneys was quitting and everything, but the next morning they said that things was starting to work again.

What happened next?

I couldn't walk. Couldn't believe I couldn't walk. Finally, I told them I wanted to come to St. Mary's for my rehab, so they sent me to St. Mary's for rehab. I got out in December. Anyway, I was in rehab for over two weeks. I still couldn't walk very good. I could walk a little but not much. I fell a couple of times at home. I got over it. The rehab people said that they thought I did really good. I made bread down at the rehab. They wanted you to do stuff like that down there.

In comparison to when you grew up, what do you think is good and bad about society today?

Well, I'd say that one thing is people can get an education if they want it. It's good that it's available like it is. Anybody that really wants to work can get an education if they want to.

There's no use to mention anything that's not good. I can think of several things that's not good. Of course, my family all are in church and most of my grandkids. I don't know if you'd call that society or not, but it's good to have your family in church with you . . . [One bad thing] about society is the way the churches are going now. Seems like they're getting more worldly. 'Course I know not all churches are like ours, but that's just the old people talking, I guess. I just don't like the way they dress . . . Not that many years ago, you never saw a person in the choir without a dress on, and 90 percent of the men had on a suit and a tie. 'Course, if you don't wear your best to church, where are you goin' to wear it? Of course, I don't ever wear a suit and a tie now, hardly ever. Not very many people do.

With religious perspectives in mind, do you have any advice for future generations?

The only advice I'd give 'em is, "Get right with the Lord. Serve the Lord so you don't have to worry about dying."

One final question, what do you like best about being from Arkansas?

Well, this is where Jimmie Lou is.

CHAPTER 4

Missing Jack

Biographical Information

Name: Carole Presley
Place of Birth: Flora, Illinois
Date of Birth: December 21, 1926
Interviewed By: Pat Temple and Chad Brewer

Where were you born?

I was born in Flora, Illinois, in 1926 . . . That's where I graduated high school from, all of us. My parents had their hands full, but we were one of the lucky families. My daddy was a railroad engineer and made good money for that time. He worked for the railroad for 40 years, and he retired from the railroad. He made good money, but with 12 kids, there wasn't a lot of extra money. We had a nice home and plenty to eat. My mama and daddy made sure we had nice clothes and shoes. We all, except my oldest sister and my oldest brother, wore a lot of hand-me-downs. We got some new stuff, but we got a lot of the hand-me-downs too! We had fun. We walked to school every day. When school was out, we would walk home and the first thing we did was change clothes. Mama made us put on old work clothes. That's what she called them. We would hang up our school clothes if they weren't dirty. Mama said washing them to much wore them out faster. We would do our chores. The boys did the outside work, like tend to the animals and do the yard work. All of us girls took care of the housework. My main job was ironing. My mama believed in ironing everything! I ironed handkerchiefs, tablecloths, and all the linens . . . After we did our chores, we would go outside and play ball or jump board. Have you ever played jump board?

No, what is it?

My dad cut a log about three feet long, and what we did was [we] took a long board and lay it across the log. A person gets on each end, and one jumps and throws the other one up in the air. When he comes down on the board, it throws the other one up! It was a lot of fun, but we had a lot of sore bottoms too . . . We had a lot of fun. Sometimes, after supper, we would play board games, but most of the time we listened to the radio. There were some good radio shows on. They even had soap operas on the radio. My daddy liked to listen to the country music shows, like the Grand Ole Opry. Well, I guess we all liked to listen to that. A lot of times, all of us kids would get up and dance. If Daddy wasn't too tired, he and Mama would dance too. You know, [they] didn't get a TV until I was grown and had moved away.

What are some of your favorite memories of your parents?

Our country drives. Like I said before, we were lucky. Not many people had a car back then, but my daddy bought a brand new 1926 Dodge the year I was born. I remember it looked like a box! He kept that car for years before he finally traded and got another one. Anyhow, on weekends when Daddy was off work, or on Sundays after church, we would all pile in that car . . . Some of us had to sit on laps! Daddy would head for the country. He taught almost all of us to drive on old country back roads, but I remember we would sing just about the whole time we were on our drives. My mama couldn't carry a tune in a bucket but she sang the loudest!

When did you move to Russellville?

The end of '46, it was right before I turned 20. I was working for the International Shoe Company in Flora. They opened a place here in Russellville, so I transferred. I was young and wanted to see what it was like to be on my own. Believe me, Mama threw a fit. She did not want her baby to get that far away from her, but thank the Lord for my daddy, he stepped in and talked her into letting me go without a fuss. I did the bookkeeping for the company . . . When I got to Russellville, I took some courses at [Arkansas] Tech in accounting and bookkeeping. That really helped me a lot. I ended up getting a good pay raise. [I met my husband] here in Russellville. I met him in a café uptown. He was about the best looking man I had ever seen. Jack was real athletic. He played football for Tech; he was good at it too. Later on Jack even coached football. [We married] in 1949. Jack and I were married for 49 years when he died. I'm 83 years old now, so Jack and I would have been married 61 years this year. I've really missed him. Jack was a good man, a hard worker. Jack worked for Tech too. He was the Director of Grounds . . . He was in the Army until after World War II, and then he joined the National Guard, but [we] had our own business. On the weekends he didn't have guard, we sold antiques. We went to antique auctions, and then we would set up in different towns and sell our antiques. We made pretty good money doing that, but we also had a lot of fun. I met a lot of different people and made a lot of good friends during those times. I miss those days.

How long have you lived in this house?

Jack and I bought it in 1965. It's a nice area . . . In the back is a Mexican family, but they are real nice, and they are clean too. They keep their yard real nice. I love this place. Jack died right out there in the back yard.

What happened to him?

That morning he had a doctor's appointment, because he had been having chest pains. He had stents put in a few years before. Anyhow, the doctor said he was setting him up for some tests the following week. I was mad! I told Jack that he didn't need to wait that long, he needed the tests now. Jack said for me to hush, he would have them done the next week. Well, later on that evening, he walked out back to do something and fell dead of a heart attack right there, right in the back yard.

Do you have Children?

Jack and I have a son that is 50, and a daughter that's 53. My daughter lives north of Dover, and she has three kids. [Her] twin girls are a lot of fun. My son doesn't have any children of his own, but he has some stepchildren. I have five great-grandkids with one on the way. The first thing [the grandkids] do when they come [to my house] is head straight for the refrigerator to see what I have cooked! My friends say their kids all do the same thing. I guess that's just a typical thing for family to do at Grandmother's house . . . My grandkids all love me! [My grandson] is a preacher at Dover for the Church of Christ.

What are some of your favorite things to do?

Besides spending time with my family, I like to work in my yard. I've been trying to get my grass to spread; it's a little bare in some places—not enough rain this year. I also like to embroider. My friends and I do that a lot. Here are linens I bought [at] a garage sale to embroidery on. I only paid five dollars for them, and I'm going to rework all of them . . . You've seen my little dogs. I love to take care of them. Well maybe they take care of me! They are a lot of company to me—all three of them. Did you notice one has only three legs? He got out of the fence, and this girl ran over him. She was talking on one of those cell phones. I didn't think [the dog] was going to make it, but he did. The vet took care of him, and said he would be fine. He is. You see him out there in the yard playing with the other two? He runs as fast on three legs as the other two do on four.

What do you think about the aging process?

[Being older] I can do what I want, when I want, and how I want! Sometimes I do nothing at all. I think, "Well I've wasted this day." But then I say, "So what. You know, if I want to waste my days that is my business." [I don't like] not being able to do the things I used to do— things I did when I was younger and in better health. I can't climb a ladder any more. You see those curtains? I took them down yesterday to wash them and almost didn't get them back up, because I couldn't climb up that ladder. I had to get that grabber [reaching tool] to put them up! I'm scared of falling. I have fallen, and it is so hard for me to get up. If I broke something, I would be in real trouble. Travelling, that is something that is hard on me now. When I ride very far my legs hurt me so bad. Earlier this month, my daughter and I had to make a trip to Denver. My legs hurt me so bad, but we had to go because my oldest sister died. She lived to be 100. Who wants to live to be 100? Well, if I can keep pretty good health and my mind don't go any farther; I guess it would be okay. If my mind goes, I had rather God take me on home. I don't want to be a burden to my family, and I don't want to be stuck in a nursing home. I do have days when my memory just is not right, but my friends tell me that forgetting is a part of aging, but that's all left up to the good Lord!

How has the world changed to you since you were young?

All the electrical things, you know, computers, Ipods, cell phones. I don't really even know what an Ipod is. Then there are all the different televisions. You can even take pictures with your phone now, but you know what has really changed? How fast the world is now. People don't have time to enjoy their life any more. That's sad to me.

What do you think about younger generations?

I think all the children are wonderful, but they need a lot of direction—too much sex! They use sex as a recreation now! It was a no-no in my day. It's sad, because so many babies are born without their fathers. I guess that's better than killing them. It was shocking in my day for a girl to give birth out of wedlock, but I bet there were a lot of abortions we didn't know about back then.

Are you in a relationship now?

I have a companion. He's five years younger than me. When he comes over, we play cards sometimes. He likes to play games on the TV game channel. I think everyone needs a companion.

Is there anything else you would like to tell us?

Well, maybe we should talk about God. Do you go to church? [I go to] the Methodist Church here in town . . . People need to give time to God. I believe that. He has let me have a wonderful life so far. I just think about what he still has in store for me! I get excited because I know Jack and I will be together again in God's presence. I do miss Jack, but life has been good to me and I've been happy, so I have no complaints.

Is there anything else that you want to add to this interview?

Just a touch of wisdom from the wise, keep God first in your life, take care of your health, always smile, and be happy . . . Always look for the bright side of things.

CHAPTER 5

That Red Hair Turned to Fire

Biographical Information

Name: Earl R. Marvel
Place of Birth: Hartman, Arkansas
Date of Birth: December 25, 1924
Interviewed By: Ashley Derrick and Jonathan Cody Tatum

Where did you grow up?

Hartman and Coal Hill [and] it was rough . . . We worked 'til we left home . . . I went in the Army when I was 18 and came out when I was 21, but I went to work at the [coal] mines when I was 15. I went overseas and stayed about 30 months. I was in four battles. I helped take four islands. [I] killed a man coming in a foxhole. After the war was over, they had lost my records. They never did find my records. [When I returned] they wouldn't give me my job back [at the coalmines]. I had to force 'em to knowin' I was going to Utah when I got it back and got my vacation pay. I was going to work in the mines there. We stayed off and on for seven years there. I came back and went to work construction for the rest of my life. [I did that for] 50 something years . . . My dad come from, Illinois and my mother from Carroll County [Arkansas].

Did you have any nicknames when you were growing up?

Dickeybird is what it started out at. I had an old gray mule we'd bought off of Dick Byrd, and I rode it all the time so they started calling me that. Then they dropped it to Dickey, and then to Dick. I stayed with that old mule about 'til we left the farm. I rode that mule everywhere around.

How was life on the farm?

My dad did farm 60 acres of land, and we growed cotton. One farmer, Coyer, owned the land. He would let us grow corn to feed the mule, but then we went to work on Thompson's land—Vaughn Thompson's land . . . We had to grow cotton all together . . . We sold the cotton to his dad. His brother ginned the cotton. It was four cents a pound for lent cotton. You could 'bout get as much for the cotton seed as you could for the cotton, but we had to keep the cotton seed to grow other plants the next year and feed our cows. We kept them in a bin, and we had two or four cows, and chickens and hogs. We never went hungry, but that's what we lived on. [Mother] worked all of her life from daylight 'til dark and patched clothes 'til midnight . . . When we was pickin' cotton, she patched cotton sacks 'cause we couldn't buy new ones every year. She fixed the old ones over and over. She got up at 5:00 in the morning and went to bed [late], but she was up at 5:00 the next morning. [Daddy] worked the same way. He worked every day. [He] drove five miles with a wagon team to work five miles each way up into the Coal Hill bottoms. He broke the land for the garden. We had a big garden all the time. He broke the land, plowed it up, and got it ready. [My brothers, sisters, and I], when it was warm weather, would run to the creek every time we got a chance. It was a couple miles to the creek, but we would run down there and swim and the rest of the kids around done the same thing.

What do remember about your grandparents?

My grandmother, one of 'em, when my youngest brother was born they took me to her house . . . I was close to five . . . She had poured [peanuts] on the porch floor, and that's the only thing that I can remember about either one of my grandmothers. One grandfather died a day or two after I was born, and then the other one lived 'til I was 14. I can remember him livin' in our old house down below us, but he was good to us kids. He'd do anything . . . He growed his own tobacco and twisted it up until he dropped dead with a heart attack.

Can you tell us more about your brothers or sisters?

I have five sisters. One of them died when she was young with whooping cough, and I still have one living. She's 94, well, will be in February. That's my youngest sister, and I have no brothers living. I'm the last one. I had four brothers; there were ten of us. One sister looked after us at school [and] anything happened she was gonna be there. My older brother, he didn't care one way or the other. He liked everybody, and that's the way he was . . . Then, my younger brother, he was the same way, but he had a temper like some of the rest of us. He said what he thought, and he didn't care who it was to.

Can you talk about the food that was important growing up in the Arkansas River Valley?

We killed everything we eat. We raised the hogs. We raised the beef every year. Mother canned it. She canned a thousand quarts of stuff every year. We had it in the cellar. Anything that was left, she canned it. We had a big cellar, and she filled them shelves every year. We didn't know what it was to run to the store in five minutes. We only bought sugar, flour, and coffee . . . My dad fox hunted . . . Anytime we caught an opossum, we skinned it and sold the hide. My dad liked to eat opossum and sweet potatoes, but I didn't. It was too greasy for me. I liked squirrels, because they weren't too greasy. We killed squirrels and ate 'em and wild rabbits. We had traps in the wintertime, three or four traps, and you'd put a grain of corn in front of the trap and an ear of corn in the back of the trap. We'd catch them rabbits and eat 'em. They were good, well, we thought they was back then, and we were proud to get 'em. You got 15 cents to a quarter for a possum hide.

Have you, as a hunter, noticed any changes in wildlife in the region over the last 50 years?

There are more and more deer. The first deer I ever saw killed, my dad had a double barrel shotgun, [and] a guy from Hartman borrowed that gun . . . [He] took it deer huntin', and he killed a deer with it. He had it layin' on his fender on the side of the car and brought it back up to the house . . . That was amazing for someone to get a deer, especially with a shotgun.

Where did you go to school?

[I went to] school at Hartman. One time I skipped school, and they caught me. There was two of us. Well, they were gonna whip us for it. They were gonna make us bend over the stairway in front of the whole study hall. Well, we both refused. [They] took us to the head coach about it . . . Whichever one got it first was gonna punch him in the mouth. Well, I was the one that he started at first. I punched him in the mouth. Out the door we both went, over the hill by the schoolhouse, and [to the] rock wall . . . We jumped off that rock, and I hit that with my neck. It flipped me back, slowed me up a little bit, and I hit the ground running again . . . We run across the holler, [coach] run us to the top of that hill, but he give up on us. We was too fast for him . . . That was the end of my schoolin'. [Actually], I was 14 and that is why I went to work at the mines.

What do you remember about your teacher at that time?

Our teacher was red headed, and she had tried everything on me. [She] put me in the corner. That didn't work. [She] put me in the cloakroom where we hung coats, and that was right down my alley. I sat there and went to sleep. Next thing, she put my nose in the ring at the blackboard. Well, when she'd turn her head, I was back showing the kids what I thought. Then, she decided to put me under her desk, which was the biggest mistake she ever made . . . I pinched her on the leg. She jerked me out from under that desk and that red hair turned to fire. When she got through, she said, "You get in that seat and you sit there and be quiet!" I did. Another teacher before her was named Hopper. She was Indian, and she was a good teacher, but one day I got to callin' her Miss Grasshopper . . . I found out she didn't like it, and she whipped me for it.

Did you stop calling her Miss Grasshopper after that?

I called her Miss Grasshopper again, and she whipped me again. Well, I had to try it again, the third time I called her Miss Grasshopper she turned the paddleboard up edgeways and used it that way. Well, my sister I was talkin 'bout was down in the study hall, and she heard about it, and she hit that room . . . She told that woman what she thought, and what she'd do to her if that ever happened again. She meant it too, but it didn't happen again, but [after that] I didn't call her Miss Grasshopper again either. It was just being ornery is all it was because she was a good teacher.

Do you have any other school stories?

I boxed at school. We had boxing, and I started as a kid boxing and liked it because it was rough. I never did play basketball or nothing, 'cause we boxed . . . I liked it. I could win. Another thing I could do, only thing I was ever good at, was arithmetic. They could give me a list of numbers, and by the time I got to the bottom I could write the answer to it . . . We didn't get to start school 'til along maybe December 'cause you started picking cotton 'bout the first of September . . . My brothers and sisters did make it through school, but I was too ornery. *Black Beauty* was the only book that I liked to read, but as far as reading and spelling, I wasn't interested.

Do you remember your first girlfriend?

My brother and his girlfriend, they were older than us. I was 14 and [my girlfriend] was 13. We went to Ozark, and they let them into a café where they sold beer . . . We'd sit out in the car and wait . . . We got married when she was 17, and I was 18. That was the only girlfriend I ever had.

Do you remember the first movie that you saw?

Yeah at Hartman, it was a Roy Rogers movie. Of course, we didn't have money to get in all of us. We'd take up whatever it was, a nickel or dime, and we'd send one in, and he'd let a rope down at the backdoor up to the second floor . . . We'd go in at the backdoor, and if they caught us coming in they'd put us out. We'd just go back around . . . The most remembered movie was *Gone with the Wind* (1939). That was after I got grown, and I had a car then, and mother always wanted to see it . . . We [had to push the car to start it]. I can remember [mother] pushing on that car 'cause she was a young woman back then. She died at 59, worked herself to death.

Do you have any favorite movies, TV shows, or actors?

I like all westerns. That's my favorite . . . I still like to watch Lucille Ball. They still show her movies every once in a while. I watch *Bonanza* every day. That's about all I do throughout the day is watch TV. That, and when this election is over, I don't know what we're going to watch . . .'Cause we're fast going into dictatorship right now.

Well then, let's wrap this up with your opinion of President Obama.

I know what dictatorship is [and] that's exactly what this [president] is leading us to, [because] anybody that knows dictatorship can see every line that he has drawn. He's sending our money overseas trying to buy them countries off. You don't buy them off, you've got to show them who's boss.

CHAPTER 6

It's Not Just a Painting

Biographical Information

Name: Elmer Dean Williams
Place of Birth: West Fork, Arkansas
Date of Birth: December 17, 1937
Interviewed By: Ashleigh Lovelace, Monica Nill, and Lindsey Virden

Have you always lived in Arkansas?

No. I was taken to California when I was six years old by my parents, and we lived there until 1991. We came back on vacation, and then in 1998 I retired and moved back to Arkansas permanently. [I moved back] because my wife said she wanted to retire here, because all the lawns looked like golf courses.

Was she from Arkansas?

No, she was from Colorado, [but she knew about Arkansas] because we came out here on vacation every year. She really enjoyed it.

You grew up north of the Arkansas River Valley in West Fork . . . What was your childhood like?

I knew that I was loved, and I knew that my parents made me feel important. [I had] six brothers and sisters . . . I'm the oldest. [Growing up] I had to milk the cows, I had to feed the chickens, and I had to take care of the garbage. I had to take the garbage out and mow the lawn, and they said I went to bed on time, which I didn't like to do at that time. [My bedtime] was 10:00.

What memories do you have of your parents?

My mother was always the best. When I talked back to her I was in a lot of trouble. My father, well, he sort of stayed to himself. He didn't talk to me sometimes, but he took me places sometime. He was a pretty busy man and worked hard to make a living.

Can you discuss your education?

I have my full education plus 2 years of college. I went to Modesto Junior College in Modesto, California and Merced Junior College in Merced, California. [I studied] psychology and electrician classes. I went to school for hydraulics.

What was your very first job?

My very first job was picking grapes, cutting grapes with a knife . . . That's my first job that I remember . . . I remember working in some lady's yard, and she gave me money for doing that . . . Then, I helped build the San Luis Dam in the 1960s in California. John F. Kennedy turned up for the first shovel of dirt. I was there when John F. Kennedy turned the first shovel of dirt . . . I didn't get to talk to him personally. He had bodyguards around him. I saw him and watched him put the shovel in the ground. Then, from there I took a job in industrial maintenance and not only worked on steam turbines, but electric machinery, worked in the boiler room . . . I was there for 30 years at that one job. I just enjoyed working. I couldn't wait to get to work. Not because anything was wrong at home, but I just wanted to be there to see what was going on. Some people hate going to work, I wanted to be there early, and I was usually the last one to leave . . . It wasn't dangerous, because I was repairing [turbines] and they would always safety lock the machine I was using. I was the only one that had a key to it so it was pretty safe . . . I still work even though I'm retired. I do electrical work and painting. I take care of my house; remodel the bathrooms and the kitchen and new floors in the house . . . That's about what I do.

How many children do you have?

I have five and one stepson—six total.

Do you have any special memories from your life you would like to discuss?

Yes, when I got my first bicycle in 1952. It was a black Schwinn. It was the new bike on the market . . . If you had one of those, you thought you were great. [Friends] were envious. Here's a story—my three-year-old grandson wrecked my truck . . . I was at my brother's house, and I put the truck in park, turned the key off, but I made the mistake of leaving the key in the car. He turned the key on and got the car out of park at three years old! He was in his diaper. The truck started rolling backwards down the hill, and when he went by me all I saw were these two big brown eyes saying, "Somebody help me! Somebody help me!" The truck ran into the bank and I had split seats in the front, and he fell through the seats and landed in the back seat. I got to him and got the truck stopped, and I said, "Dalton you wrecked my truck," and he said, 'It's okay papa, I'll buy you a new one. What color do you want?" I think I have ten [grandchildren] plus five great-grandchildren.

Tell us a little about your wife.

[She] is a very energetic person. I call her a workaholic. She works at St. Mary's hospital in Russellville, Arkansas. [She] is a secretary, among a lot of other things. Before that, she was a cosmetologist for 34 years. [I've been married to] Retta Jean for 44 years. I'm eight years older, or maybe its nine. We get along good, and we don't fight or anything. We just try to compromise. If she wants to do something and I don't, I'll do it anyways. If I want to do something and she don't, she'll do it anyways. We kind of know one another well enough that we know how to get along, you know? Naturally, we have watched other people throughout life . . . Some people don't get along at all, and sometimes I don't know how they stay together.

You will be 73 this year. Do you feel like your age is catching up with you?

Well, at 72 I don't feel like I'm 72. I haven't had very many health problems, and I don't take any kind of medication at all . . . My doctor says that I am in good health. So, I don't worry about anything. I'm real energetic, and I love to do things. I love to work. I don't like to sit at home and twiddle my fingers. I like to be active and doing something. The reason I'm like this is because after I finish a project that I'm working on it has to be just right. Everything has to be just right . . . The people I'm working for right now, painting a house inside and out, told my wife that they don't want anyone else to paint their house because I'm a professional. [Keeping busy] helps me physically and mentally . . . I read a lot. I read books that my granddaughter gets me. I read all kinds of books—medical books, health books, um, Jane Roberts' books, Louis L'Amour books. I don't like all books. Most books that I like I can become part of the story in the book. I can visualize it. Some of it I know it's not true [while] some of it makes you stop and think about . . . It could be that way, and I could be there as part of it. Does that make sense? I'm not much of a hunter, but I do like fishing. I'm an usher at the church, and I like that. I'm a greeter at the door, and I enjoy that . . . I like cars. I've built some old classic cars in my time.

Let's talk about those cars.

I built three cars. [One was] a 1936 Ford pickup. I paid $1,000 for it and sold it for about 12 times that. I took the body off of it. I put an all new undercarriage under it 'cause in those days they only had two elliptical springs—one in the front and one in the back. When I got through with it, it was like a brand new truck. It had power steering, mustang suspension, water gauges, and a 305 Chevy small block engine in it . . . I was invited to join the street club, but I wanted to stay independent . . . I enjoyed just my wife and I driving it up in the mountains . . . I showed it at the fair, and it won first place. I sold a '54 Cadillac Coupe de Ville I built. I painted it black. I paid $400 for it, restored it, [and] sold it for a lot more. I won't say how much, but lots more.

Can you discuss family traditions?

We've always had family reunions for as long as I can remember. Eight or nine, I can remember those. It was fun. You get to see everybody, your aunts and uncles, and cousins, and everything—people that you haven't seen. Our immediate family and intermediate family was real close . . . [We would] all go fishing sometimes. Good times . . . We celebrate all holidays. On Christmas and Thanksgiving, family comes over. We have a big dinner, sometimes music and karaoke. You know, things that you just like to do . . . We play football and baseball out in the yard, and horseshoes. Those things are kind of traditions for us . . . I've enjoyed doing that. [Holidays] are commercialized a lot, but other than that, I love Christmas and Thanksgiving.

What are the advantages of being older?

I have time to read more, time to study more. I feel that these things can keep you mentally sharp, and I just like to know what's going on. I read everything, the newspaper and even the junk mail sometimes . . . I just love life. I just love it. I love everything around me. I love all the trees, and I know they are all alive . . . I just love creation, but I don't only just think about it, I enjoy the creation. I enjoy knowing that there was a creator. A more super person that could create these things . . . Like, if I look at a painting, most people just look at the painting. They don't think about the artist that created the painting or visualize what he was doing to make it look so beautiful . . . It's not just a painting. It's a lot more than that . . . The earth is not just a creation. There was a creator, so that's what I like. I know that I feel like even the atmospheres that we have are alive, the trees are alive, the grass is alive, and the squirrels . . . I know that the creator created that for me to enjoy.

That is an interesting perspective. What else can you say about aging?

I feel free. If I don't want to work, I'm not pressured that I have to go to work . . . When I do go to work, I stay 'til, the job [is finished]. The job I'm on now has been four weeks. Some of them are longer . . . I enjoy knowing that I know how to do the things that I do. I figure that it's kind of special to me. I am my own person . . . I feel just relaxed most of the time . . . If I don't want to go out to dinner, I won't go out to dinner. If I want to go shopping, I only shop for special things that I like . . . [Actually], I can't find anything that I don't like about aging . . . I don't want to live forever, but I want to live as long as I can. Knowing that I can hold onto feeling like I am important and I know that I am [important] to someone, I'm thankful for everyday.

Have you had any significant health problems?

Well, when I was 62, I had a major heart attack . . . [After that], I had a shoebox full of medicine they gave me. I eventually threw all of [it] away. I had an aneurism operation three years ago that was successful . . . I got a letter from a doctor that I am in perfect health. I don't have any problems . . . I take vitamins, juices, and enjoy life . . . I don't require very much sleep. At 5:00, I'm up every morning and at 10:00 that night I go to bed. Well, that part is a routine because when I worked I got up at 3:00 every morning, when I had my job I retired from . . . I feel what makes you get old is getting stressed out, and you have to learn to not do that. [You can't] worry about things you have no control over. Worry about the things you do have control over.

You mention working a lot. Do you think that working helps you feel young?

Yes . . . To me it's very uncomfortable to sit around and not have anything to do. Some people just sit around. They don't go nowhere. They don't do nothing. They don't go watch the fall colors when they come in. They don't watch the trees as they change the colors . . . These things are real important to me . . .

What else helps you feel young?

I have a dog, and she really is a blessing to me. [We] have quality time together. She is very smart and intelligent for a dog. I know what she wants and she knows what I want . . . She's always glad to see me. I pull up in the driveway in the car, and she doesn't know how to get to the garage. She runs to the door and tries to get to the door, and runs to the garage door and waits for me to come through. I think that everybody needs a pet, because pets they don't expect a lot from you. They want to be loved, and they want to love you back. Things like that, I think that's what makes the world go round.

CHAPTER 7

I Grew Weed

Biographical Information

Name: Georgia J. Loyd
Place of Birth: Booneville, Arkansas
Date of Birth: October 17, 1938
Interviewed By: Kyle L. Smith and Brianna Hennington

Do you remember the first house you lived in?

I don't . . . I was so young when we moved from there, but I remember our home was located four miles north of Booneville, Arkansas in the small township of Chismville. We moved there in 1947, and our home only had four rooms in it . . . Our land consisted of 160 acres that was primarily woods and forest. Our home had no running water or electricity. We had no bathroom in the house. Mother canned vegetables, so she always kept a pretty good size garden. We had chickens, and Mother would kill a hog every winter for us. I milked cows. In fact, I was the main cow milker. We didn't slaughter our own beef.

Can you talk about your family and growing up with them?

I had four brothers named C.E., Shelburne, Donald, and Jerry. I had one sister named Glenda. She was the baby of the family. I played a lot with my brothers, since I was basically the only girl old enough to play with them. We didn't have toys, so we made up our own games. We used to love to play baseball, and I was actually pretty good at it! My mother's name was Lillie Bryant, and my father's name was George Bryant. My mother was a stay at home worker . . . My father ended up doing church work in the latter part of his life by becoming an evangelist within the Church of Christ.

What else do you remember about him?

He was placed in the [Arkansas State] Tuberculosis Sanatorium in south Logan County when I was an infant. He contracted [tuberculosis] and was placed on what was called "The Death List." When I was three weeks old, my father looked at me through a screen door because he wasn't allowed to touch me or get close to me. Many people were institutionalized at the old sanatorium because TB was prevalent and killing many people at that time. My father surprisingly enough was released from the sanatorium, and later moved back in with us. He died when he was 65 and my mother died when she was 85.

What were some of the scariest times in your childhood?

When I was younger, I used to stay at my grandpa and grandma's house. My grandpa was diagnosed with a blood clot in his brain, and it made him into a violent person. One day, my grandmother fixed my grandfather something to eat. Grandma said he had been acting strange all morning. He was eating his food very fast and was acting angry and hostile all morning toward her. My younger sister Glenda was the only grandchild present at their home that day playing out in the yard. My grandmother walked in the back of the house momentarily to do something. When she walked back into the dining area, my grandfather pointed a double barrel shotgun at her head, and [using profanity] stated he was going to kill her. It was then my grandmother ran outside and screamed at my sister, "Glenda run! Glenda run! Papa is going to kill us!" Glenda and grandma ran as fast as they could into the woods to get away from him. My brother Shelburne and his friend pulled up at the house and grandpa was sitting outside with the shotgun inside the door way. Shelburne went in, and took the gun, and ran out the back door. We called Doctor McDougal to come check on him, and grandpa became angry. [Grandpa] ran inside for his gun, but it was gone. He then said he knew Shelburne was up to something and he was going to kill him. After that, I stayed at their home trying to tend to him and help my grandmother.

That sounds frightening. Can you think of any other scary experiences?

When I was young and still living at my grandpa and my grandma's house, I used to walk a mile to the school every weekday. I used to be so scared, because I walked by myself down that dirt road to school. There was a man named Mr. Stewart who lived on the road that I traveled to go to school. Mr. Stewart had severe migraines and was bed ridden for the majority of his life. Every day I would stop by Mr. Stewart's, and he would always have bananas and vanilla wafers for me, because I stopped and checked on him. I was always scared after I left his house because there were no other houses on that road back to my house. When I used to leave his house, I would run so fast until I would get home. One day, I was walking back from school, and just as I passed Mr. Stewart's, I saw a man down the road in the distance. I became scared, but kept walking with my head down holding onto my books. I remember being so scared I thought about cutting through the woods, but I was afraid of the cemetery [that was in the woods]. As I got to where the man was standing in the road, there was no one there. I saw this wagon pulled by horses pulled over by the side of the road. I then heard something down along the side of the road in the ditch. When I looked down in the ditch, there was a man down there who was staring at me. I knew of him and he was not a very pleasant man . . . He started laughing . . . He had very long hair and a very long beard, and was nasty and dirty looking. He kept smiling and chuckling as I walked by. I remember being so scared that I almost started to make a dash for it, but I kept walking with my head down until I was able to get over the hill from him. Once I got him out of sight, I started [running] home until I made it to the door.

Can you talk about your husband?

I met Junior in April of 1957 in Booneville. There was a little café there in town called the Palace Café . . . I was a waitress there for a little while. There was a man named Jimmy Loyd who was related to Junior . . . His fiancée at the time [Shirley] set us up on a date. Junior worked at the movie theatre in town and also raised chickens at the time. We started dating and on August 9, 1957, we got married. I was 18 and

Junior was 25 when we got married. We have been married for 53 years now. One time while me and Junior were dating, I was also working at the local Piggly Wiggly service station when Shirley pulled up in her new car. I remember telling Shirley, "Let's drive down to Blue Mountain and drive in front of Junior's." Blue Mountain was 13 miles east of us, but we decided to be rebels and make the drive . . . Junior and Jimmy were outside working on a tractor when we pulled up . . . [Attempting to act coy], we kept driving and ended up driving into a creek . . . We got stuck and soakin' wet. It was embarrassing. We had to get Jimmy to get his tractor and pull the car out of the creek! At that time, we wore poodle skirts, and my dress faded its colors on my legs. It ruined my dress, and Junior had to drive me back to work. Needless to say, I was a little late getting back to work.

What major historical events do you remember the most?

Of course, I remember most of the wars and things such as 911 [the September 11, 2001 terrorist attacks] . . . The most memorable was when John F. Kennedy was assassinated. I was at my in-laws residence when it came on the television. It was probably the first national historical event that really captivated me. I remember feeling funny knowing at that particular time in history that we didn't have a president . . . Of course, Johnson was sworn in swiftly, but I remember feeling odd that I didn't have a president at that time.

What are your biggest educational achievements?

I did complete high school when I was younger. I made it all the way through the 12th grade. I also attended Draughons [Practical Business College] in Fort Smith. I was not able to finish the courses, because my mom and dad ran out of money to pay the bill.

Can you talk about jobs you've had?

Well, I worked at a hospital for a few years. I did clerk work and paperwork for the hospital, but I didn't stay very long. I also helped my mother manage a laundromat she owned for a while as well. As far as a career, I worked for the Booneville Water Department for 22 years. I miss the social interaction at the office and the feeling of a close-knit family we had—that's what I miss most about working at the water office in Booneville. I have always enjoyed gardening and growing plants. One time at the office, I found some seeds outside, and I planted them in a vase and watered them for several months. I set the plant in the drive-through window at the office where people drive by and pay their bills, because it was such a neat plant. One day, a man came in the office and told me I might want to take the marijuana plant out of the drive-through window. I grew weed accidently at the office! I still laugh about that 'til this day, and so do my former co-workers.

You aren't working now, so do you enjoy retirement?

I enjoy being retired . . . My main source of joy from being retired is getting to spend more time being with my family. I really enjoy getting to watch my grandchildren play sports and [seeing] their school functions. I really loved sewing, but my hands don't let me do much of it anymore . . . I mainly love being able to travel a little more to Oklahoma to see my daughter Kim and her children. I love to work puzzles. I believe they keep my mind sharp. I do have a lot of time on my hands, but I stay busy keeping the house clean and helping my husband do things. I also enjoy cooking for my family. Almost every Sunday after church, I cook for my children and grandchildren. I am now a great-grandmother as well. One of my granddaughters and her husband live in Clarksville, and they just had a new baby boy.

What do you miss the most about the "old days?"

The thing I miss most about the good ole days is the simplicity of life. Everything back then just seemed a whole lot simpler. Back then, we didn't listen to the news and hear about all the bad things in the world. I'm sure bad things were going on, but you didn't hear about them day in and day out. Another aspect of life back then that I miss is the family togetherness. Families back then just seemed closer and did more things together. I remember almost every Sunday afternoon, all the kids in the area used to come to our house to play in the yard—a bunch of kids! We had a blast and enjoyed being around one another and socializing with one another. I always said I could climb a tree just as high as any boy could! Back then, you made your own fun, not just playing video games.

I just miss that aspect of life back then and definitely can tell times are totally different.

Is there anything you like about the world now?

The things I like best about the age in which we live now is the progression of medical advancements and the more improved medical science out there. Although the news is depressing and is frustrating sometimes, it is nice to be able to keep up with world events with the new technologies. I also really like the fact that children, for the most part, have better educational opportunities. Back then, most kids dropped out [of school] before they could graduate. I really like the cell phone. The cell phone is another part of the modern age that is really useful and helpful. It makes it easier on keeping up with your loved ones and checking on them when they are away.

Wrapping things up, is there anything else you would like to add to this interview?

I have had a great life and great family. Like I said, I enjoy being around my kids and grandkids. I haven't had a very exciting life, but I have had a very blessed life . . . I love my family.

CHAPTER 8

The Ice Sign

Biographical Information

Name: Gerry Crabtree
Date of Birth: August 1, 1937
Place of Birth: Mountain Pine, Arkansas
Interviewed by: Bo Jordan and Landon Walker

Where were you born?

[I was born in] Mountain Pine . . . a logging town that was owned by the Dierks family. The Dierks owned Mountain Pine and everything in it. So, when we moved there, [Dierks Lumber and Coal Company] gave our family an allotted amount of money at the grocery store, job security for anyone who wanted to work within Mountain Pine, and so on and so forth. Families were encouraged to shop at specific stores, go to specific movie theaters, and work at specific locations because it meant the money was being put back into the town. So, my mother and father moved there for the benefits and security after the depression.

What was your home life like as a child?

Wonderful . . . We lived simply, and my mother and father were both wonderful parents . . . If we needed milk and butter, Mother would tell me to go down to the creek . . . We had a wire cage that [we put milk and butter in that] sat in the cold creek water. Daddy had a smoke house that he smoked ham, turkey, deer, [and] different meats. I remember one time, Daddy scolded me after being in the smoke house, and cutting the crust off the outside of the hams that hung in there, and eating it with my friends. It wasn't until [Daddy] passed that my [husband] explained to me that once I had cut the outside of the ham off, it exposed the inside of the ham, and the flies would get into it and cause the meat to go bad.

Can you talk some more about your parents?

My dad was pretty much a big kid. He played with us and teased us. He was an excellent provider and parent with high morals . . . He also knew when to discipline, but that was primarily my mother's job. Daddy was a carpenter. He helped build lots of the buildings in Hot Springs. My mother, she ran the house, the money, and was adamant about education. She raised me and my sisters to be ladies—to be Southern . . . She never went anywhere without dressing as if she were going into town. Even in the house, she washed dishes in nylons, dress, face of make-up, jewelry—everything. Whether she was in the garden or washing clothes, there was never a time when she wasn't dressed up enough to go into town. She was a lady above everything else, and her children were to follow suit. She made our clothes, was strict, and was a highly moral person. If we were ever sick, my mother always treated us with home remedies. [She] occasionally drank, mainly for medicinal purposes . . . When she found out us kids were smoking, she decided she would start. The thing was; it took her a week to smoke one cigarette. She would light the cigarette, take a drag or two off of it, and put it in the window sill in the kitchen.

Do you have any siblings?

There were four of us girls. My oldest sister Donna Jean died when she was six of diphtheria [an upper respiratory illness]. The next sister is Bonnie, who was the one right above me . . . We had washing machines that had rollers like a conveyor system, and you fed the clothes through the rollers to wring all the water out of them before hanging them on the line. When we were little, I put Bonnie's hair in it. Then there's me, and then the youngest was Glenda, who died in her 40s from arthritis.

What is your fondest childhood memory?

My favorite memory is of being asked by Mother to place a sign in the window of our house to let the ice man know how much we needed. The sign had the numbers 10, 20, 40 and 50 printed on it . . . We would circle how many pounds of ice we wanted with a piece of chalk, and we'd place the sign in the window. The ice was brought to us in solid blocks. The ice man would pull the block of ice out of the back of the ice truck with what looked like a big logging clamp. The clamp had big teeth in it to grip the ice block. After he took the ice out of the truck, he would take it to the back of the house and place it in the icebox. The iceboxes were not very well insulated back then, so we would have to go out back every couple of hours and dump the water from the pan underneath it.

How would you describe the clothing from when you were growing up?

Well, I made most of my clothes. Girls didn't wear pants very often back then. We wore skirts. I made all of my skirts at home, and the few pair of pants that I did have were bought at a store. The skirts were usually calf length and straight or a-lined, and the pants were usually cropped like Capri pants are today. I wore white poplin [button down] shirts usually, with a cashmere sweater over the top. Back then, a cashmere sweater only cost $25. If you go to buy one today, you would pay at least $125. I always took very good care of my clothes, which was probably because if I wanted anything special I had to pay for it myself. My older sister often took my sweaters and wore them. I would get so mad at her, because she did not take care of them the way that I did.

Did you always live in Mountain Pine?

Well let's see. We didn't stay in Mountain Pine long, we moved to Hot Springs where [Daddy] got a better paying job, and this was right after [World War II] had started. When we were in Hot Springs, we had a black nanny. That was normal back then. After we left Hot Springs, we moved to Oregon and owned a large farm. [Daddy] grew hops, potatoes, several fruit trees, and he also had a berry field. Back then, the farms weren't just hundreds of acres; they were thousands of acres. Mother and Daddy hired several Filipino farmhands to help tend the fields, but Daddy was always working too. He never tried to put the work off on anyone else. During this time, every man was being drafted and shipped off to war. The best way to avoid being drafted and deployed oversees was to own a farm that provided for the government. We left Oregon when I was in 6th grade and moved back to Arkansas. That was shortly after the war ended. I will never forget [the day the war ended]. It was so loud. School was let out early that day, and there were whistles and sirens going off everywhere. When we moved back to Hot Springs, Daddy helped build Second Baptist Church . . . After that, we moved to California where I went to high school. I stayed there until after my kids were born, and we moved back to Arkansas when I was in my 40s.

What was your first job?

My first job was working for a veterinarian off Highway 7 in Hot Springs. I was 12 and in the 6th grade. They hired me to make dog food. I was paid $10 per week, [and] that felt like a lot of money especially during that time. Do you want to know what I bought with my first paycheck? Do you know those old '50s style metal lawn chairs? Well, I bought a two person slider version of one of those for Mother. She put it on the front porch. My second job was at a clothing store called Melody Dress Shop. I lied and told the store that I was 16, when I was really only 13. I worked there for 75 cents per hour. When I left there, I started working across the street at another clothing store where I got paid $1 per hour. After that job, I started working at a dry cleaner at the end of the street for $1.25. You know, when you are really young, you are going to work for whoever will pay you the most money . . . I worked at Nance Chevrolet when I got older as a switchboard operator.

It seems like your family didn't hurt for money, so why did you have so many jobs growing up?

Well, my older sister never worked . . . She always got what she wanted, but if I wanted anything beyond the necessities, I had to pay for it myself. When I was 16, I really wanted a car. Back then, you did not have to have insurance in order to buy a car. Daddy told me that he would [give me a down payment on] a car, but I had to make all of the payments on it. My first car was a '36 Ford Coupe with a trundle seat in the back. Getting that car and having to pay for it helped shape me into the woman that I am today. I was always driven, but getting that car taught me to always push for anything that I wanted.

You've been married more than once, right? Can you talk about your first husband?

He was a financial provider, but he was not good for the family. I was still married to him when we moved back to Arkansas. We were divorced when my oldest was a sophomore in high school. When I divorced him, he divorced the kids. We were married in California and had four kids— the oldest is Donna Jean. She's my only girl. Next, is Darrell Wayne, and he is just 10 months younger than Donna. Then, there is Dwayne Allen, who is two years younger than Darrell. Then, there is the baby Leo, who is two years younger than Dwayne. Donna is married to Rick, and they both work at the Wal-Mart Home Office in Bentonville. They have one daughter named Toni . . . Darrell is married to Karen. He is a physician's assistant, and she is a 3rd grade teacher. They have two kids named Chase and Kelli. Dwayne is married to Melissa, and they work at a metal fabrication shop. They have two kids named James and Jessie . . . Leo is married to Denise, and they work at the metal fabrication shop as well. They have one son named Josh and one granddaughter named Cadence. I am blessed to have all of my children and grandchildren live within two hours of me.

How did you meet your second husband?

Well, I worked at Ideal Bread Company in Hot Springs . . . This little man kept coming in and telling me there is someone he wanted me to meet. I was not very interested at the time, but one day, the man that he had wanted me to meet came in the store. He was there to buy barrel bread, which was the three-day-old bread that was collected from the stores. They used it for feed for the animals. He asked me to go to coffee, and I agreed. We dated for one year before we were married on October 15, 1977. He is the best dad [to my kids] and the best husband I could have ever asked for . . . He is my best friend.

Do you have any family traditions?

Every year [my sons] wrap the front porch in [plastic sheeting], and [my husband] gets the wood burning stove ready. All of the kids and grandkids come to the house on Christmas Eve, and we put ornaments on the Christmas tree . . . We play dirty Santa [a game where family members draw numbers for gifts and have the ability to steal gifts from others depending on the number drawn], start the bonfire, and pass out Christmas pajamas to all of the grandkids. Everybody stays the night at our house, and all the grandkids sleep outside on the porch. On Christmas morning, I make quiche, and then we all sit on the porch and open presents. At noon, we eat Christmas lunch . . . Every year in the summer, the entire family goes on vacation together. Usually we go to the beach, but this past year we went to Gatlinburg, Tennessee. On family vacation, each of the kids and their individual families has one night that they are responsible for cooking dinner. We always pick one night to do a crab and shrimp boil.

How do you want to be remembered?

I want to be remembered for being a good mother and for being a good provider for my children and grandchildren.

CHAPTER 9

No

Biographical Information

Name: Irma May Sherman
Place of Birth: Scottsville, Arkansas
Date of Birth: August 28, 1932
Interviewed By: Kayla Osborne, Whitley McCormic, Yesenia Lopez

Where did you grow up?

[I grew up in] Scottsville. [I worked] on a farm [and went to school] at Scottsville from kindergarten to 6th grade and to Hector from 7th grade to 12th. [School] was boring. [I remember] Mrs. Linda and Mr. Tom. They were nice teachers. I knew them all my life. Mrs. Linda taught the first six grades and Mr. Tom taught the other six. [My favorite subject] was spelling.

How many generations has your family lived in this region?

Three or four I guess. Now that was on my daddy's side. They came from Tennessee and Missouri. I don't know about my mama's.

What do you remember about your parents and grandparents?

Daddy was always out working in the field. He cut firewood, bailed hay, made sorghum molasses, tended to the cows. [We farmed] corn, peas, taters, beans, strawberries, blackberries, [and] cucumbers . . . I learned how to milk a cow when I was five and how to pick cotton at five! Mama always stayed at the house. She took care of the house, and the kids, and the garden. She canned vegetables and churned butter. We sold the butter and eggs we got. Mama quilted and taught me how. My mom's dad was a doctor. A medical doctor, that is. They lived the next town over, about eight miles away from us. My dad's parents were dead before I was born, so I don't remember them.

What do you remember about growing up in Scottsville?

We were all home for [family meals]. We'd sit down and all eat at the same time. [We went to church] every Sunday and Sunday night. On Sunday afternoon, [we would play] horseshoes and softball. If it was a rainin', we'd play pick-up sticks or hide the thimble [hide-and-seek with objects]. We'd play checkers, and fox and geese [a tag game]. That's all we had. I mean, the kids today, don't know how to play and have fun. We made our own playhouses. We'd get out and take rock, and build the outline of the house. We'd get a long rock, and that'd be our car. You just pretended a lot. We didn't have electricity 'til I was around 16 or 18. We had kerosene lamps. That was all we had, so we thought we could see pretty good, but boy once we got them electric lights, we sure didn't like for them to go out! [Our first car] was a Studebaker. [I didn't graduate high school because] my daddy broke his leg. [He] busted his kneecap out, and I needed to help on the farm. My sisters helped Mama, but I'd rather be out [farming]. I hated housework. I hated to clean—still do.

What was your favorite holiday growing up?

Christmas . . . It just came once a year . . . [I would get] one doll, maybe a box of handkerchiefs, candy, apples and oranges. There was a community Christmas tree that would have filled this whole room. There was icicles on the tree and that was about it, and handkerchiefs on it. Dolls, if the parents could afford dolls . . . Now, kids have Christmas every day. You see something you want; you go get it, or you con someone else into getting it for ya. Christmas meant something to us because it was that one time of the year. There were people who had more money than we did, and they got to do more stuff . . . I was satisfied with what I had.

Do you remember your first job?

Hauling hay . . . [I] drove a truck in the hay fields [and] hauled square bales of hay . . . I would help unload them, [but] I didn't help load them, because I drove the truck.

Do you have any brothers and sisters? Are they still living?

Well, there was five of us girls, and we each had two brothers. Only thing is, each of us girls had the same two brothers . . . [The girls] are still alive and both boys died . . . Gene was in his 50s or 60s and Willis was in his 80s.

You were married. Can you tell us about your husband?

I probably [met him] at church. I don't remember . . . That was too long ago . . . [He was the one] and I just knew it! [We married] in 1954 [and he died] in 1999.

Do you have any advice for future generations?

Just always be honest [and] do right.

Is there anything in life you did not do that you wished you had?

No.

CHAPTER 10

Get Your Sideburns Cut or Get a Guitar

Biographical Information

Name: James C. Wheeler
Place of Birth: Clarksville, Arkansas
Date of Birth: June 11, 1938
Interviewed By: Alesha Charles

Can you tell me a little about yourself?

I'm James C. Wheeler . . . I go by the initials J.C. I was born and raised, for the most part, here in Clarksville, Arkansas. I was gone for five or six years during my childhood, but for the most part I was here in Clarksville. My parents are Adrith Orvil Wheeler [and] Etta Louise Basham. She's 93 now. She's still with us, but she's in a nursing home. She's doing good, feisty and fighting all the time . . . They got married here in Clarksville, but I do not know how they met . . . in Clarksville.

How many siblings do you have?

I have one brother and one sister. I had one other brother that was a miscarriage in 1946 I think. I never was told about it, but I found out about it. I don't know why they kept it from us but they did . . . My brother is Robert Arthur, and my sister is Janice Lee. She's a Hanshaw. She married Hanshaw . . . 50 something years ago I guess. They had six boys and one girl . . . My brother, I think he was named after my two grandfathers Robert and Arthur . . . I was named, as the story goes anyway, after a friend of my dad's named Jake Walton. He used to live right down here in Clarksville.

What do you remember about your childhood?

Well, we did quite a bit of camping and fishing when my dad was home. Dad was a truck driver, so he was gone a lot. Mom did most of the bringing up, but he did what he could with the job he had. He drove a truck up until he was 75. At Christmas, we always celebrated Christmas real big. We had a local gentleman here that would play Santa Claus. They would always sneak what few gifts they had for us outside, and then he would come in the front door with them . . . One Christmas, I think I was probably five, Santa Claus asked me my name, and I said J.C. Wheeler. [Then], he asked my brother his name, and he was scared to death. He said J.C. Wheeler. He gave the same answer I gave!

Did you have any favorite toys?

Maybe my first bicycle . . . It was a 26 inch. I was about six and couldn't even touch the pedals. We didn't have the stuff they have today. We took an old pet milk can and played kick the can with it. [We] took a steel iron/hub thing off a wagon wheel and nailed a tin can to the end of the stick and bent it like a hoop, and we would roll that steel ring. That was our toys. We didn't have nothing like they got today. [Kids today] get too many [toys] . . . They expect them, but they don't appreciate them. We played baseball in an area that wasn't any bigger than this room. We played a lot of cowboys and Indians. Over where the cemetery is now, Oakland Cemetery, the western part of that over by Harvest Foods [grocery store], that was all pastures and wooded over in there. That is where we did most of our playing.

Can you talk about technological advancements throughout your life?

When I was born, I was born in my grandmother Wheeler's brand new house. They didn't have electric lights. They had gas lights. Gas pipes came down, and you lit the gas lamp hanging from the ceiling. The telephone was a big wooden box with a crank on it. You held an earphone up and talked into the mouthpiece. There wasn't a dial. There was no phone numbers. You just crank the thing like Radar O' Reilly does on [the television show] M*A*S*H* and tell them who you want to talk to . . . My first car I traded a Schwinn bicycle and a five-dollar bill for. [It was] a 1937 Ford. There wasn't nothing on it but a stick shift, and a steering wheel, and brake pedal. There was no power anything. The first car that I had that had any power on it [was] a 1956 Cadillac Sedan Deville in 1962. It had all the new features on it—power steering, power brakes, air conditioning. It was the first one I ever owned that had any of the new stuff on it. I never had one that had GPS on it. I had a '29 Hudson that I bought in '65. I give $325 for it. I put a new starter, new six-volt battery in it, and five gallons of gasoline . . . [It had] six wooden spoke wheels, four on the ground, and one on each front fender for spares. I would love to have that one back today. It was built like a tank.

Did you have a black and white television?

Oh yeah, that's all we had. There was a dance show . . . I never danced. I never did learn to dance. I never wanted to dance. I never had any interest in it, but I used to watch [dance shows] a little bit . . . I remember the first TV we got. We lived in Sherwood, [Arkansas]. That was something else. I believe that was 1950 . . . By today's standards, it was crude and wasn't put together real well, but it was exciting to see something on television.

Did you ever serve in the military?

I joined the Air Force [a few] days after I turned 17. I wouldn't advise kids to do today what I did then, because there is so much change in the world . . . When I was in 10th grade in 1955, I told our principal, Johnny Bock, that I wouldn't be back the next year. I was going in the Air Force. He said, "No, you'll be back." I did come back, but I was in uniform when I came back to see him in September of '55. I had been planning for about two years to go to the Air Force and make a career of it. I turned 17 on Saturday [and] was in Little Rock on Monday . . . I was in basic training on Tuesday in San Antonio, Texas, with the intent of making master sergeant. At that time, that was the highest you could go in the enlisted rank and retire with 20 years. I made master sergeant in 15 years and retired with exactly 20 years in 1975. I re-enlisted the last enlistment, so I re-enlisted for the last four years. As soon as I took my hand down, I turned around and said, "Now where do I put my retirement papers in?" They said, "Are you sure you want to do this now?" I said, "I knew in 1955 when I was going to do this." I put my retirement papers in the same day I enlisted for the last time. I stayed 20 years, and it was planned out that way. I did have some real fine gentlemen during my career that helped me get in the right career field and keep pointed where I wanted to go. They actually had more to do with it than my parents, because they were there and knew what it was going to take. The first gentleman I went to work for was Master Sergeant William J. Nicholson at Kelly Air Force Base. He took me under his wing like a turkey would a young one. He guided me for about the first five years I was in the service . . . He kept me straight . . . About that time Elvis Presley was going real good. In the military you don't wear long side burns [like mine]. They would start getting a little long, and he would tell me, "Speedy," that was the nickname he gave me, "Speedy, you're going to have to get your side burns cut or get you a guitar." This went on for some time. One morning, right after roll call, I knew he was about ready to tell me to get a haircut . . . I went over to him I said, "Nick, you got $10 I can borrow?" He said, "What do you need $10 for?" I said, "I'm going to buy me a guitar." Whew, Whew, Whew . . . I thought he was going to screw his head in the ceiling, but he got over it and got a good laugh about it. The next morning I showed up with a new haircut.

What was your job in the military?

I [started] as aircraft maintenance. I was at my first assignment maybe six months, and I started flying with the aircraft as a flight mechanic. A lot of it was top secret. We carried our own maintenance people. We took care of our own airplane, because we couldn't have everybody from these other bases working on the airplane seeing what's in there . . . I flew over Japan, Europe, and North Africa. I would be gone anywhere from a week to sometimes as much as three weeks at a time. That was for about the first five years. Then, in 1961, I was transferred to the Azores Portuguese Islands, about 800 miles off the coast of Spain and Portugal. I spent two years there . . . I got a back injury that took me out of the maintenance end of it and moved me into maintenance control. We coordinated the maintenance between the aircraft [and personnel] as they came through . . . In '62, I came out of the Azores to Wichita Falls, Texas— Sheppard Air Force Base. I went to work as assistant dock chief . . . I worked there for a while and went back out on the EC-47 as a crew chief. I became stand board evaluator for the EC-47 aircraft, [which was a plane that intercepted enemy radio communications for intelligence purposes and used communication signals to pinpoint enemy locations]. Then, I moved to the T-29, [which was a training plane for bombardiers, radar operators, and navigators]. I was dual qualified. I was a flight examiner on the EC-47s. I was an instructor on the T-29s.

Can you tell me about any specific missions?

I had my T-29 out at the wash rank one day, and Gary House came down there and said, "Ms. Berry wants to see you." Ms. Berry worked in personnel, and when Ms. Berry wants to see you, you have an assignment. I cranked the aircraft up and taxied it back down to the base . . . [Ms. Berry] had an assignment to Vietnam for me, but I didn't have to take it because my wife was eight months pregnant with our youngest daughter. I had 72 hours to accept the assignment or reject it. If I rejected it, as soon as the baby was born I would have been gone and who knows [where] . . . The wife and I discussed it and opted to go ahead take it . . . The assignment was a brand new electronics, warfare mission in Vietnam. I got in on the ground floor of it. I was one of the first ones assigned to it . . . I [joined] a crew in New Hampshire in late August of '66, and we flew an EC-47 across the Pacific to Saigon. We had an extra navigator and an extra pilot for the trip. Colonel Hinkle measured the gas, and Captain Harris and myself went to Nha Trang and started the new 361st. I was supposed to have stayed at Tonsinute, but Colonel Hinkle told me the day before he was to leave, "Jim have your bags packed in the morning. I'm taking you with me." [So], I never had orders to go to Nha Trang, but I stayed there the full year . . . There were three squadrons. Each squadron had 18 to 20 airplanes and about three times that many crews. We were the only crew to ferry over, fly together, and come home together. We flew 114 seven-hour recon missions. That was the best year I had in the Air Force. It was a good crew, good airplane, and a good mission . . . We were there finding Charlie or North Vietnamese whoever. We would pick up their radio signals and do radio direction finding on them. We had equipment that could lock on to that signal. If they were at 270 degrees, they were west of you. It would point 270 degrees. We would print all of this and take three, four, or five different shots. Because we were flying, we could get different angles . . .

Some of these targets were just logged for reference for later dates. Some of them had the B-52's called in—the Arc Light Missions . . . That is the B-52's coming in and dumping loads of bombs. We set up some of the Arc Light Missions targets. We set up targets for some of the fighter planes to come in and take them out. That is what we were doing . . . I did get to fly the aircraft, and I did perform as a navigator a lot. I have a large chart that I worked in Laos, where I was actually plotting the fixes. Colonel Hinkle took that chart, unbeknownst to me, until we landed. He wrote up a certificate type thing on that chart, and the whole crew signed it. [It stated] I was authorized to fly as navigator any time that crew was in flight. That is a keepsake.

How was it to be on secret missions?

Nobody got on the airplane besides the flight crews and our own ground crews. Until I put my website [www.ec47.com] up, there was absolutely nothing on the web about the EC-47's—absolutely nothing. I started coordinating EC-47 reunions in '98 . . . I was on the phone with one of those high-powered civilians in security service, which was the ones who provided our back end crews linguist [support]. He wanted to know if I would be the spokesman for the news media at a ceremony we were going to have [in San Antonio]. I said, "I'm not a speaker, but I guess I can." He started to tell me now you're going to have to watch what you say [about EC-47 missions]. I said, "Wait a minute before you go into that. You need to go to my website and see what is out there for the world to see about what we did." The next day I get a telephone call back from him. He said, "Impressive to say the least." That was all he could say about it . . .

With everything being top secret, did you ever get to make phone calls home?

We could make a phone call occasionally through Military Affiliated Radio Service. They work with the HAM radio operators here in the states who would patch you into a telephone system. I am a HAM radio operator myself and have been for about 22 years. Yeah, we did make calls home, but since it was relayed through radio and then to the telephone, when you finished your statement you would have to say, "Over." Other than that, it was the letters and of course the little inexpensive tape recorders came out. There were a lot of guys that used those and sent little tapes opposed to a letter . . . [I never] used them, but my best friend over there he did. I caught him one night making an audio tape to send to his wife. I thought, I'm going to get him. I got my 38 [revolver], went out the front door, and came around the back. He's sitting right next to the window—no window, just a screen wire. I let three rounds off just as quick as I could pull the trigger into the ground . . . He had to make his tape over again.

Do you have any other military stories?

Down at Little Rock in '71, I was a flight engineer on the C-130s. We were in Germany on a 60-day rotation, and we left Rhineland going to Athens, Greece. We leveled off somewhere around 30,000 feet. The co-pilot gets out of his seat and goes to the back of the airplane. Well, I just unbuckled out of my seat and slid over into his seat. The aircraft commander said, "When you get comfortable let me know, and you can have it." I got comfortable, and I said, "OK, I want to hand fly it. I don't want no auto pilot." I flew that thing to Athens started flying the approach to land. I keep thinking, "He's going to tell me to get out of the seat here in a little bit." Finally, we're down about eight or ten thousand feet and he says, "Did you ever land one of these?" and I said, "No sir." He said, "Do you think you can?" I said, "Yes sir." He said, "You're fixing to get your chance." I landed that thing, and when I landed he was sitting there with his arms crossed. You couldn't feel it touch the runway. It was a beautiful landing. After we landed, the co-pilot [who had left] said, "I'm a qualified co-pilot and never made a landing like that." I just kind of turned my head around and said, "I sure would hate to have to duplicate it." We had 17 passengers, and we weighed 137,000 pounds. [Letting me land the plane] wasn't legal. Oh, it wasn't legal at all. It was my first landing in a large aircraft.

Flying sounds hard. How did you learn to fly?

In the Air Force . . . There's not that much to it as far as flying the aircraft and keeping it straight and level. You know what you want it to do, and you got to anticipate what it's doing and what it's going to do . . . The 5th of August 2007, I was taking my first [official] flying lesson right out here at the airport. [I bought an old plane], and I was restoring it. I took off and stuck that sucker in the ground like a lawn dart. I broke my back! I caught a power line. We lost power, and the aircraft wouldn't stay up. We were too far down the runway to land, and were too low to try to turn and come back. [My recovery] took a few months. I was in a hard brace from [just under the neck to above the hip] for two months. I was supposed to wear it for three, but I came out of it. That was all I could handle of that thing. I sold my plane for half of what I had in it and said, "That's all!"

How did you meet your wife?

Well, I was home on leave. We had relatives, an aunt and uncle and cousins, in Vanduser, Missouri. We were going up to visit them. I was driving. It was my car. My brother and his girlfriend were in the backseat. I was kind of flirting with her through the rear view mirror. That was in '55 or early '56. Anyway, I ended up taking his girlfriend away from him. We got married October 23, 1956. She was two days short of 16, and I had turned 18 in June. She just passed away the 11th of August. We were 2 months and 12 days short of 54 years.

Do you think when you were growing up people had different values?

Well, I just think the whole world is different today. I think the values are altogether different than they were back then. When we got married, we said until death do us part, and we meant it. I don't know too many [people I graduated with] that are still married to the same one they married some 50 years ago. There is a few, but not too many . . . I know there is a few that their spouses passed away as mine has, and they remarried. Today, it is just I want a divorce—bye.

Do you have any marriage advice?

I don't know how to give advice on it. I know what I did. I wouldn't argue with her. I would just turn around and walk off! In almost 54 years, I've never laid a hand on her, but we had some rough times. We ate an awful lot of pinto beans, turnip greens, and corn bread—an awful lot of that stuff . . . Everybody today, they have got to have the big car, and when the money gets tight they lose it all. They get into too much debt. I did the same thing. I got into debt. I never did lose anything, but I did get into debt. It was sure hard to get out of it. Let it Not be Afraid

Do you have any special memories of your children from when they were growing up?

I was flying a lot. I was like my dad in a truck. He was gone a lot, and I was gone a lot. There was two years I was gone for a year at a time. The year I spent in New Jersey because I couldn't get housing for my family, and the year I spent in Vietnam. That was two years I was gone, completely gone for a year at a time . . . My son who was the second one born, he's 13 months younger than my oldest daughter. We were in San Antonio, and my mother and father were down there because they were there to help with [my daughter] while my wife was in the hospital with the new baby. About midnight, the wife said, "I need to go." So, I run Kathie over to my mom and dad's and I came back and picked Betty up and headed for Wilford Hall Hospital out at Lackland. I ran red lights and everything straight out there. There wasn't any traffic. I get out there and they met me at the first floor . . . They took her to the delivery room on the third floor. I went back to the first floor to register at the hospital. I wasn't down there about two minutes . . . When the door opened when I got back, they said, "You've got a baby boy."

At that time, could men be in the delivery room when their child was born?

I don't know if it was allowed or not. I wasn't in there for any of them. I waited in the waiting room.

Is parenting different today?

Well, you can't do it today, but parenting for me was, "Do like your supposed to, or you're going to get a paddling." Now, I'm not going to whoop one of my grandkids. I would if the parent agreed, but I'm not going to go against their wishes. The parents today, there is just so much more that they have to look out for than what we had to look out for. There is all kinds of drugs and thugs, and so many more things they can get into today. Back when my kids were growing up, we didn't worry about marijuana. We didn't worry about drugs. There wasn't that much of it out there. That is just one part of it, but you know what I'm saying. The world is changing so fast, so fast.

Have you done everything you wanted to do in your life?

I've done pretty well everything I wanted to do. There is a couple of places I would like to visit again. I'd like to go back to the Azores and visit. In fact, I've got a daughter who is a cancer survivor and it has been three years or so ago, I was going to take her back over there for her birthday. She was born there in 1961. I was going to take her back over there for her birthday, but she has so much trouble traveling . . . I'm going to try to go over there myself and spend a few days. I would like to anyway. I'd like to go back to Vietnam again and visit there. I would like to go to the villa that I lived in. It's still there. I've got a photo of it that was taken about four years ago. I've got the photo from the front that I took in 1966. I've got a friend over there in Saigon he's teaching and writing. He went to Nha Trang and stood where I stood to take my picture, and he took a picture. The house is still there, but there is no yard. I mean, the buildings are right up against it now, but the house itself still looks the same. I get a lot of photos of Vietnam now. If I didn't know it was Vietnam, I wouldn't believe it. [It looks] like downtown New York.

What has been the most fulfilling thing for you?

It would be a toss-up between marriage and my military career. The marriage was great. I can see it now [since my wife died] a lot better than I could see it before. There's a lot of things I wished I had given a little more thought to, and talked about a little more, before she passed away. I talk to her and tell her things now. I talk to her photo. I don't think she can hear me, but she might be able to. I don't know if I'll ever remarry. I doubt it, but I might . . . I know where I'm going to be buried, right next to my wife.

CHAPTER 11

Still Alive and That's About It

Biographical Information

Name: Kenneth Wayne Hottinger
Place of Birth: Dover, Arkansas
Date of Birth: June 3, 1937
Interviewed By: Bailee Miller and Tarren Roberts

Where were you born?

[I was born] in Dover here in Pope County. That old house right up on top of the hill. After I graduated from high school, I went to California for six months. I worked at the post office until I got enough money to come back home . . . [My childhood was] real good. I mean, we were poor, but we had whatever we wanted I guess . . . I had one brother—Charles Kirby. I played baseball.

What was it like when you were growing up?

The grade school that I went to was on the square. It's a big two or three story building on the square where the supermarket is now . . . They tore it down [and built a new one] on the hill where the [middle] school is now, or pretty close to where it is now. Our [high] school burnt when I was in the 10th grade [then we] used the old agricultural building . . . Everybody [was] packed in it for about a year. My dad died in 1957, my last year in high school. My uncle and aunt [Earl and Annie

Shinn] lived down here below us. I bought a bunch of cows, milked them in their dairy farm, and sold milk. I drove a school bus. I was the first senior in high school to drive a school bus . . . Just a high school education is all I got. I graduated from Dover.

What did your parents do?

My mom was just a housekeeper, [and] my dad had a lumber truck. He hauled lumber, and they had a saw mill . . . He and my grandpa, they just sawed lumber and planed it and carried it off.

Can you talk about your wife?

We've been married 47 years . . . [I met Charlotte] in Malvern, Arkansas. I worked at Burke Equipment Company. It was a parts store, and she worked at a restaurant—Ballard's. I'd stop there and eat breakfast and lunch, and I met her there . . . [We dated] six months and got married . . . She had just gotten out of high school. We married November 16th in 1962. I had that wrote down [for this interview] . . . I still can't remember. I can't remember birthdays or nothing. I used to could a little bit better than I can now.

Didn't you start your own business here in the River Valley?

[Charlotte and I] started out with a truck wash in Pottsville [in 1986] . . . We had it for 13 years I guess . . . We sold it [in 1996], and me and my son went into industrial services [with a company called K-Hott]. We've been in that ever since.

What are some of your favorite things to do?

I rabbit hunt, and I deer hunt. I got my beagles that I [American Kennel Club] field trial with . . . I've been doing it about 25 or 30 years. I still work a little and relax when I want to, sleep as late as I want to . . . I do just about what I want to do. [I started it because] I had the beagles to rabbit hunt with, and I had AKC registered beagles. So, I got a club started in Russellville [called] the River Valley Beagle Club. [With field trials], you have a 13-inch then a 15-inch, and a male or a female. Then you just turn a pack loose of seven. You got two judges that are qualified to judge, and they run along with the beagles and see what they do—how fast, checks, and all that you know. It's kind of interesting, but it's hard to explain . . . We've got one next Saturday and Sunday at Booneville. I've won a few, [and] I had a few field champions . . . I try to watch all college football games. I enjoy them. I'm a Razorback fan.

Can you talk about your family?

Christmas and Thanksgiving is our biggest holidays. We enjoy all get-togethers at my daughter's house . . . My brother's dead. He died about 15 years ago. It's just me and my wife, and [my son] Kenny and his wife Tina, [and grandchildren] Wayne, Jessica, and Isabella. I've outlived all my aunts and uncles. They're all gone . . . Me and Kenny went to Canada once [to rabbit hunt].

Is the world different now from when you were younger?

I was six years old when we got electricity for the first time . . . We had a one-room school [by the cemetery], and I went to the 6th grade there. We rode bicycles or walked . . . We didn't have no computers in school or nothing like that. That didn't come along until I got out. I can remember I was in the 10th grade at Dover, and Mr. Reed our Algebra teacher got a TV . . . For Halloween, he invited us all down there to watch TV for the first time. You just barely could see it [because it was] too fuzzy. We've had a lot of changes. Cars, everything's a lot different.

What was your first vehicle?

I had a '49 Ford. Used to, when we was dating or anything, we would drive our dad's car. [Everyone had] one car per family if you're lucky. Not every kid had a car, but I did own a '49 Ford. I guess that was in '55 or '56—something like that. [Since then], I had a Chevrolet, a Toyota, a Chevrolet, a Dodge, and now I got a Ford . . . I guess I'm not particular.

Are there events that stand out in your mind that relate to war, politics, or entertainment?

I can remember when World War II was over . . . I was in the 1st grade [when] I came home from school, and they said it was over . . . I actually didn't know too much about it. Folks always talked about it . . . The Vietnam War was terrible . . . The boys were over and fought, and when they came back people called them "baby killers" and all that . . . They went to protect their country, and they were doing what they thought was right or at least what they thought they were supposed to be doing . . . I don't think they got a very good check over it or anything like that, but people thought they were right in protesting it. It wasn't the boys' fault. They had to go . . . I missed it some way or another, and I don't know if I was too old I can't remember just how it was. Well, I joined the National Guard, and I just got out of it I guess . . . The worst disaster was the 9/11 [terrorist attacks on New York and Washington D.C. in 2001]. That was terrible. I can't think of anything else really that was as terrible. [With politics], I really liked Ronald Reagan. He was a good president. [With entertainment], we always thought Elvis was mighty fine.

What is your opinion of younger generations?

They're real good and smart. They sure are. They're something to be proud of I think. They just, they're goin' to make it fine—most of them.

Would you consider your aging experience to be a positive or negative one?

Positive. I think you get smarter . . . You still get smarter when you're 73. Things come to you more. Things will happen that you didn't think it would. You'll learn more, but you never get too old to learn . . . I can't get around as good as I used to I guess. Can't do the things I used to do, but I [enjoy] retirement and got a check coming every month. That's a good part of it . . . I'm still alive, and anyway that's about it.

CHAPTER 12

We Would All Turn Over

Biographical Information

Name: Martha Henderson
Place of Birth: Shelbyville, Tennessee
Date of Birth: May 5, 1935
Interviewed By: Maggie Treece and Catherine Wilkins

How was it living in Shelbyville?

I don't know . . . My father was a minister [and] we moved around quite a bit. The first place I can remember, when I was big enough to remember, was Anniston, Alabama. He pastored a [Church of God] in Anniston. I [lived there] about two years until I was about two or three, maybe four. I had a real good childhood . . . I had a brother and a sister. My sister was 16 years older than I am, and my brother was 13 years older. I was the baby. I [also] have a brother that is living, and he is four years older than I am. There was four of us.

When did you move to Arkansas?

We moved to Arkansas in 1940, when I was only five. It was around the time of the Great Depression, [and] you had to have coupons or something for food or to buy groceries with. Whenever we first moved from Big Sandy, Tennessee, we lived in London, Arkansas. We lived there only about until my brother started school, and he only went to school there about one year. Then, we moved to Augusta, [Arkansas]. I lived there for quite a few years. I started school there. I quit in the 10th grade, and then I went back and got my GED.

Besides your dad being a preacher, what do you remember about your parents?

My mother was a quiet person. She didn't have a lot to say, but my dad . . . We sang together. We did a lot of things together. Then, we had to pick cotton, and chop cotton, do things like that. I always went with him to do that. It was really a good time that we were living in [even though] it was kind of hard making it. I know we had some good times. My dad was a very loving person. I can remember that he would, even after I was grown, [let me] sit in his lap . . . He would rock me. Whenever I was younger, I would get scared 'cause I would think I would hear someone outside my window, and my mom would always have to come sleep with me. They really had patience with me, even after I got pretty good size. I didn't like to sleep by myself.

Did having a preacher for a father influence your religious views?

As far back as I can remember, when I was a little girl, I always wanted to go to heaven. So, being a Christian was a great desire. I wanted to be a Christian and a good Christian. [With my religion] we had lots of meetings, state meetings and general meetings, and conventions. My mom and dad always made sure that I went. Growing up, that was mostly my entertainment I guess . . . Even now, I like to go to any kinds of meetings, a lot of conferences, and things like that. [I am still active] in church.

Did it seem like you had so many siblings since some of them were so much older?

My sister husband's went off and left her with three little kids when I was little. Her oldest one was only two years younger than I am . . . She came back home, because she didn't have nowhere else to go. She came and lived with us . . . We would all sleep together. Back then, you know you didn't have real big houses. You didn't have your own bed or anything . . . When one of us would turn over, it was just a full size bed and there were five of us, we all would turn over . . . [My sister] lived with us for quite some time . . . She always seemed like a mother to me [and] her kids were brothers and sisters instead of nieces and nephews.

What memories from childhood stand out to you?

[I remember a tornado] in Judsonia. It come through Augusta and Bald Knob too. In fact, I think it wiped Judsonia out. I forgot what year it was. I think it was in the '50s. My [oldest] brother lived over at Bald Knob, and it blew his house well off the foundation and set it down on the ground. From one bedroom window to the next bedroom window, it just laid the end off . . . It did so much damage over there, and I remember my dad coming up. He was over in the West Helena area, and he came home and just cried and cried.

Let's talk about music.

I wasn't an Elvis fan. Conway Twitty's first wife was my youngest brother's girlfriend's sister. [Conway] used to come over to Augusta. I really didn't know him, but I went to school with Billy Ray Smith. Do you know him? He was a big football player, and he played for the [University of Arkansas] Razorbacks. Went on to pro, and he is from Augusta!

Let's talk about your husband.

We have known each other all of our lives. We have known each other probably since we were about 10. Yes, [we grew up together] . . . He is very helpful. He always wants to do things for me, but sometimes I like to do things on my own . . . He is always wanting to help somebody, no matter who it is, he is wanting to help people. We have known each other over 50 years . . . He lived in Ohio [when we first met]. He would come down [to Augusta to visit], and him and his nephew and me and my girlfriend, we all would go on dates together . . . Whenever we got married, my father had moved to Peoria, Illinois. He had been pastoring a church there. We moved from Peoria to Toledo, Ohio. That was where all my children were born . . . We lived there probably 15 years. Then, we moved back to Augusta for a short period, and then my husband got a job in Benton at Alcoa. We lived there for about 25 years. I think it was about 25 years before we moved here. [We moved back to the River

Valley] whenever [my son] Kenny and [his wife] Vickie moved back here. We retired, and they wanted us to move.

Talk about your current family.

I have four, three boys and one girl . . . I have nine grandchildren, and I have one great-grand daughter. I really enjoyed my children when they were little, but of course, they became teenagers. It was difficult, but I made it through it . . . I really enjoy my grandchildren too. They're very special to me. I always think that family is the most important thing a person has . . . We go a lot of places together—even the children and the grandchildren. Every year, we have a family reunion for all my children, and my sister's children, and my brother's family . . . We have had it different places. We have had it in Hot Springs for many years, and then we went to Jonesboro for probably two years, and last year we had it here. We had it over at Kenny's [in] the barn . . . It turned out really nice. It is really nice that we can all get together. It seems like I am on the go all the time . . . We have taken cruises. We took two cruises to the Bahamas . . . Then, we went on an Alaskan cruise probably four or five years ago. That was really nice. I have a brother that lives in North Carolina, and we would see him a lot.

Did you work when you were raising your children?

No. Whenever my children were little, I didn't work. My husband worked two jobs, and I stayed at home and took care of the kids. I really didn't work very much until my youngest one was a senior in high school. I started back to work in retail. I worked in retail for several years. Then whenever I moved to Russellville, I went to work for the school in the cafeteria. I loved it. I worked nine years after we moved here.

You worked in a cafeteria, so you must like to cook.

I like to cook. I used to like to cook better than I do now. I like to have family in for dinner and cook for them and things like that. I used to quilt a lot and crochet too. I have arthritis in my hand. It makes it hard to quilt really. [I have] a lot of aches and pains, but I feel like I'm blessed. I can still go . . . I had to get a knee replacement, but I got over it real fast. I really feel like I'm blessed with my health.

What is your opinion of the younger generation?

Well I think that some are very, very nice, and others I would want to put a finger to . . . Really, I think that our world is changing. Teenagers today, and young adults today, are a lot different from whenever I grew up. Even the change is different from when my kids were teenagers, but I think that so many of them have good values . . . I feel like God is a part of our lives, and when I see people that love God, that makes a big difference in what they do, and in where they go and how they act.

Do you have any words of wisdom to pass down?

I think that you have to live your life as if [it might] end today. We don't have no promise of tomorrow. We must plan things, but know that even if our life ends, we have to set our goals on things higher than this world.

CHAPTER 13

My Chevrolet Was Better Than His Ford

Biographical Information

Name: Martha Williams
Place of Birth: Dyer, Arkansas
Date of Birth: December 23, 1916
Interviewed By: Alyssa Jones, Kelsey George, and Jennifer Hinojos

Tell us about growing up in Arkansas.

I was born in Dyer. My ancestors were Dyers as well, and they formed the little town of Dyer. Although I was born [there], we moved early to Alma . . . It's pretty close to my heart . . . I grew up in Alma, and there were five in my family. I had two brothers and two sisters. I was the middle child . . . My father and brother are both deceased, and one sister [is alive], but she's had a stroke . . . I'm about the only one really left in my family.

What do you remember about school?

I did have one brother that wasn't very interested in school. I don't think we would have ever gotten him through high school if we hadn't [completed] his workbooks and special projects . . . He just stated that he wasn't going to do it. We girls did the note booking and the reports so he could make it through. My parents were very interested in us getting a good education. I know my father pushed me real hard to make real good grades. He [and his brother] had been to the University of Arkansas in 1901, but he would never tell me that he had been a student there. I later figured it out though. It was because he didn't complete his work, but he could not tell me that because he wanted me to finish my work and be a good student . . . My home was only one block away from Alma school . . . I would walk to school every day . . . My girls only lived three or four blocks away from Russellville High School, but we had to take them in a car . . . They just didn't want to walk. I was educated at Alma all 12 years. The little town of Alma had a lot to do with my rearing. I am very thankful for that background.

What do you remember about your high school days?

I was in the English club, and I also played basketball. I wasn't a very good player, but I had some teammates that were good. I enjoyed playing very much. They have a position called a side-center. My husband used to say that's what I was. I would say, "No I'm not, I'm a forward." The rivalry between Alma, Mulberry, and Van Buren was quite something.

Did you attend college?

After I graduated high school, I went to Fayetteville. I had a cousin who was principal at Alma High School, and I liked what I saw in him and I thought, well if he [went to college], then I can too . . . I had a scholarship, but it was only for $35 . . . They had a program [that] paid students 18 cents an hour . . . The [Dean of Men] was in charge of the scholarship . . . [He] didn't want me to have a scholarship, because I was female. He believed that those scholarships should all go to men . . . I had a job in the educational library getting materials ready for teachers who were coming to teach in the summer. That job helped me a lot. My parents also gave me $25. I'm sure things are quite different today. I had some wonderful instructors, and I am very grateful for their guidance and suggestions. I earned a bachelor's in Home Economics and a master's from the University of Arkansas.

Can you talk a little about your family?

My husband had been to [Arkansas] Tech, and a professor there knew a professor at George Washington University in Washington D.C. He called and got my husband a scholarship to play baseball there. [After he] finished his law degree, he came back home [to Danville] . . . I went to Danville as a home demonstration agent in 1941. We became good friends. [We] married [the same year]. We had two children together. One is now a history teacher, and the other is a psychologist. She does evaluations of students at Wake Forest Hospital in South Carolina. She also teaches a class there. After we had the two children, I didn't work until they had started 1st grade.

Do you have grandchildren?

I have grandchildren. My oldest grandson graduated from Tyler High School in Texas. He had 600 students in his class, and he was number one. He is on the staff at Texas A&M . . . He will also be going to Germany in November to take his equipment and research . . . It's about protein research . . . He's a sweet kid and is so unassuming. He can just take a book, go out to the mountains, and be happy. His sister is the firecracker. She is a dancer and a singer, and is very good at both. She actually went to New York, but then decided that wasn't the life that she wanted. So, she called her parents to give her $600 to take a paralegal course [then] decided that she would like to be a lawyer. She then went to Denver to a school where they were dealing with children and families . . . That's what she's doing now . . . My daughter that is a psychologist has a son. He just moved from Kansas City to St. Louis, Missouri. He works in business.

After working as a home demonstration agent, what did you do?

I was a home economics teacher at Russellville Junior High School, but I also filled in for geography, science, and civics classes. [I then] worked at Arkansas Tech. Mr. Hull was the president at the time, and so he called to find out what I needed to do to get qualified for counseling in psychology. I read all of the theories. It's an interesting subject. I think that all my friends thought I was losing my mind because I would drive to Conway three times a week [for counseling certification]. I believe it was one of the best things I've ever done, because it helped me understand myself, and others, better than anything I could have done [through] my life experiences. Eventually, [I became] the Dean of Women at Arkansas Tech.

What did your husband end up doing?

My husband was very politically inclined. He first ran for prosecuting attorney after he returned from World War II. We worked very hard to get him elected . . . He hadn't practiced law any at all that summer, so he says, "We'll have to borrow money to eat on since we don't have any money coming in." I told him, "No, I'm not going to borrow money to eat on. I'm going to go to work." That's when I started teaching [and eventually] went to Arkansas Tech. [My husband] was also a senator and a judge in 1956.

How would you assess your life with him?

We had a wonderful life together . . . One of my daughters is just like him, but the other is different. He died here at Wildflower [assisted living in Russellville] in October. I had a room here too, because I needed to be close to him . . . After he died, I just stayed.

Was any particular time of year, like a holiday, special for you two?

Christmas was something, but since my husband has died I don't like to [talk about] Christmas.

We can skip over holiday questions then. Did you and your husband travel much?

[We] had an RV and saw everything in the states up to Alaska . . . We've [traveled to] England and all through Europe. We did this during the war days. My husband did a lot during the war . . . We went through all the command posts, and it was a chance to [see many places].

You have mentioned WWII twice. Aside from traveling, what did you do during the war?

Pearl Harbor was December 7, 1941. [My husband and I] were married the following Sunday. All the family came, and all the relatives supported us. We all knew that because he was young, he would be going to war. We would live two months in one place, he would be trained in [another] area, and then we would be transferred somewhere else. Luckily, I had a Chevrolet, and he had a Ford . . . My Chevrolet was better than his Ford. We used that Ford as our savings account! When we got to where we had to start using money [to fix the broken down] Ford, it meant that I had to find work, so that's what I did.

Where were you living then?

[We were] in Trenton, New Jersey. I had an engineering class [in college], so [a company there] hired me to examine spare parts of airplanes . . . When there wasn't any work to do, I would try to go home. I was raised that if you weren't working, then it was wrong to take money. [My boss] said, "No you can't [go home] because we need you to be here when the work comes." That was an eye opener for me. I didn't realize that you [could not] do anything and still get paid.

What is your opinion of younger generations?

Oh, I think they are all are great.

Does that include the "hippie" generation?

Oh, [that] was just something they were going through. Someone would say something about the long hair on the boys in church, and I would reply, "It's just a fad they're going through. Let them enjoy it." Every age has had that, whether it's a mustache, long hair, or something else. It's like, even now girls have certain things they wear now . . . [It's] just a fad.

We are going to shift into some aging questions. How is your health?

So many people here have ailments and aren't able to walk. [They] have walkers or [oxygen] tanks. I'm almost embarrassed as I try to walk around them, because I have good health . . . I've taken care of my body. I know what nutrition is, and I know that you need exercise . . . I have no ailments [and do not have to live here], but I didn't need to stay in my house by myself . . . I put it up for sale. The sooner it sells the better.

So, what is it like to live in an assisted living facility?

You're safe. You don't have to cook or clean up anything. They also have a list of things you can substitute if you don't like what they have on the menu. I find that the cottage cheese and fruit and the things they offer is ample for me most of the time, but a lot of other people here [complain]. I guess it's just a fact of life, people complain about food. When I was the Dean of Women at Tech, I had a lot of complaints about food . . . I have a recliner and a small table [in my apartment here]. I walk a lot every day. My main concern is to get my walking in. I am a walker.

What are the best and worst parts of aging?

[The best part] is that I have no ailments. I don't have to see a doctor or anything . . . Wrinkles are the worst part . . . I wish I had skin like I did when I was younger, but I know that's not possible. I don't worry about it . . . I see it as, each day is just a day, and you live for something in each one.

CHAPTER 14

Way to Go Howdy Doody

Biographical Information

Name: John Douglas Singleton
Place of Birth: Clarendon, Arkansas
Date of Birth: April 13, 1943
Interviewed by: Haley Meeks and Casey Dabney

Tell us about growing up in Arkansas.

My family has lived in the Ozarks for three generations . . . [Though I was] born in Clarendon, I spent most of my life in Atkins. I grew up in a poor family. My father hunted animals for food . . . There was very little work, but my father fished commercially and guided hunters. My mother was a good cook, very loving, and took care of her children very well. We were always clean, even though all we had as far as plumbing was a faucet in the kitchen for washing dishes . . . We never had a bathtub, and that's how we lived for the first 12 to 14 years of my life. My father was a good man, a very hard worker even when there wasn't any work around. There was a button factory when we grew up. They made buttons out of shells . . . We would go to the creek and gather shells, but soon it was shut down because of the invention of plastic . . . People didn't need buttons made out of shells anymore. My dad probably made $20 to $25 a week at the button factory, which was plenty enough to live off of back then. My mother had nine kids. After the fourth, she had a miscarriage, so there was a pretty big gap between the first four and the last five. It's almost like there was two different sets of kids because of the age gap. The last two were a set of twins. I have two older brothers, three younger brothers, one older sister, and two younger sisters. Three [of my] brothers are now deceased, [and] all my sisters are still alive. Their names are Glenda, Silvia, Regina, and [my] brothers names are Jeff, Avey, Paul Wayne, Jerry, and Terry.

What is your earliest childhood memory?

My earliest childhood memory was when my mother broke a plate of beans over my dad's head! We played outside . . . We played around the house, mostly games that did not involve anything that had to be bought. In the summer, we played outside a lot. During that time, children didn't stay in the house. They played outside all day . . . We also played basketball, horseshoes, and baseball. We also played a game called "Annie Over" where kids got on both sides of the house and threw the ball over. The idea was to catch the ball, and that took a lot of the day. We made these things called bean flips. We made them out of tire inner tubes and a forked stick. We would shoot rocks and beets out of it. We would fly kites, and sometimes make them. We were never bored or had nothing to do. [Sometimes I watched] shows like *Mickey Mouse Club* and *Howdy Doody*.

What do you remember about home life?

I grew up in a small, wood frame house. [It had] no insulation, so it got very cold in the winter . . . We had old tin stoves that when they got hot they turned cherry red, and the fire would go out in the middle of the night. So, someone would have to get [up] and start the stove. Me and my siblings had our moments . . . We would tie into each other. They had to sleep multiple beds—boys with boys and girls with girls. The girls helped Mom with washing dishes and cleaning the house, boys and girls helped skin animals, and the boys would chop wood.

Did you have any nicknames in school?

My nickname was similar to my dad's. His was "Peaches" and mine was "Little Peach." [I am] not sure how that came about. When in school, we wore mostly hand me downs . . . The poor kids got out of school to pick cotton . . . My favorite subject was English. I really enjoyed the poetry aspect the most. My favorite teacher was Mrs. Eason in the

6th grade. I was in love with her. I finished high school, and my class went to Washington, D.C. for our senior trip. I didn't go, because I had already signed up for the Navy. I went directly to San Diego, California, and became a sailor . . . I was in the Navy for six and a half years. Then, I went to college at Arkansas State University . . . My grandkids go to Arkansas Tech. I have a grandson that graduated May 2012. I have never heard anything bad about Arkansas Tech.

What are some other things you remember about growing up?

Family meals were mostly vegetables and fish from the river. We ate a lot of squirrel and a lot of deer. Sometimes, we would eat rabbit and raccoon. Raccoon was normally barbequed and served with yams. It was one of my favorites. We ate a lot of dried beans, boiled, and lots of potatoes . . . For breakfast, we would have gravy and sausage and sweet tea to drink. I remember sometimes we would come in from a movie, and the only thing we would have to eat would be ketchup and bread . . . We would have us a ketchup sandwich . . . I grew up in the age of the automobile, but we never had a family car. Neither my mother nor father could drive. We never had a car 'til we got in our late teens or after we got in the military. [I also remember] my first kiss with a neighbor named Gracie. That's as far as I'll go on that one.

Would you call her your first girlfriend?

No. Later on when I was in the Navy, I met a Jewish girl from Long Island, New York. We dated for six months while she was going to college, and we never participated in any activity other than kissing and stuff . . . Her being Jewish and me being Christian, I believe her family had some influence about it. [I never] saw her again after she went back home. Yeah, I did not date until I got into the Navy—never took a girl to the movies or nothing. That had a lot to do with the lack of money. Besides that, I had low self-esteem. I was introverted, and I didn't feel like I was what the girls were looking for. After I got in the Navy, I got over my shyness.

What was your favorite holiday when you were young?

My favorite holiday was Christmas. It was a time of peace and goodwill. It had not yet developed in the material aspect it has now.

Well, materialism at Christmas did exist then, and our parents somehow managed to buy a few things for us . . . We had very good food. We had wild duck, and my favorite was the breast of the wild duck. It is a dark meat with a gamey taste—very delicious. My favorite part was the Christmas parade. When I was probably about nine, we had went to the parade and the fire trucks, and what not, would throw candy out to the crowd. I ran about a mile during that parade and gathered as much candy as I could. By the time we left, I was loaded down with candy . . . They took a picture, and it was in the paper . . . I was just a rag-a-muffin of a kid loaded down, and I looked very greedy. I think my mama was glad to see all that candy. It was kind of funny. It would have been embarrassing, but I didn't know what embarrassing was. My mother did though, but she enjoyed the candy.

How did your family obtain food?

We had a [vegetable] garden and raised chickens. [We traded seeds] with neighbors. As far as meat, my dad killed all the meat we consumed. [My family] does not hunt regularly anymore. If they do, they do eat a portion of what they kill. [We grew] 90 percent of our food on our own when I was growing up. My family does continue to grow a garden in the summer, but we buy more than we ever have before. [Now], the fish are contaminated, and the deer are way more populated.

Speaking of environmental related issues, has your family been affected by natural gas drilling?

I'd say since my wife got a gas lease on her farm, yes and positively.

As you moved on into your teen years, tell us what did you do for fun?

Well, of course, we had school activities as a teenager. We all had our friends. Even though we were poor, we always had friends over to visit us . . . We went roller skating and bowling . . . My sisters were cheerleaders and [my brothers and I] would play football . . . We were small, so we sat on the bench a lot. I remember pretty much all my fellow students that I went to high school with, because we were a small school of 30 or 40 students in my class. My closest friend was probably Basil Baker. [He] went on to work at Arkansas Nuclear One and retired after

30 years. He is now deceased from cancer . . . [The first movie I saw at the theater] was with Gene Autry and Roy Rogers. I saw *The Robe* in the '50s . . . It had to do with Jesus Christ and his crucifixion.

Can you tell us a Navy story?

I served in the military from 1961 to 1967. I was trained in electronic battery-driven torpedoes. I also was trained in nuclear weapons. I learned to scuba dive. I went to Vietnam in a submarine in 1965 and mined a harbor. There is no record, and none of my service shows up on my military records because the missions we were on were too secret. My military service was good . . . Once, I had worked in a torpedo shop in Pearl Harbor on a submarine base on the Island of Oahu. We went swimming at a waterfall that was about 50 or 60 feet and maybe 75 feet in another section. My friends for some reason had got to calling me Howdy Doody . . . I don't know why, because I don't think I look like Howdy Doody. Anyhow, I get to the top of this real high ledge, and I really didn't think I wanted to go. I knew if I didn't they would call me chicken . . . One of my friends hollered up, "If you jump, we won't ever call you Howdy Doody again!" I said, "Alright." Finally, instead of jumping, I dove off and turned sideways up in the air. I blacked out before I hit bottom . . . As soon as I came up from the surface, my friend said, very loudly, "Way to go Howdy Doody!" The thing about it is, another young man from our torpedo shop jumped and hit his head on the rocks . . . [He] didn't make it.

How many times have you been out of the state of Arkansas?

I have been out of the state a few times—six years for the military and about ten just working out of state. Other people, if they don't know any better, they think Arkansas is just a hillbilly state. [I would never live] anywhere else. [When I was gone], my heart never left—never really thought about making any other state my home.

You mentioned your wife earlier. Tell us about your wife and kids.

I have been married one time. It started out as a double date. My sister was married to my wife's brother, and we decided we were right for each other. We got married in my wife's family living room by the justice of the peace. As far as I know, it was legal, and I am not going to double check it. At any rate, it was in her living room. I thought it was a pretty serious time, but seems like the majority of the people gathered around thought it was a pretty funny thing. They got a lot of laughs out of it, and I still haven't figured it out to this day . . . I hope the joke is not on me. My wife was a farm girl, and she grew up and [was] raised like me. First thing I did after we got married was I took her to town and got her a pair of shoes . . . We have four of the best kids in the whole world—until the grandchildren came along. We are a family that has survived, because we have a great depth of love about us. [My four] children, Jenny, Becky, James, and Lawrence, they are and always will be my best friends . . . I have had the good fortune of working with all four of my children at some point in time, and that was a wonderful experience. I have so many memories about my children growing up. The boys would fight like boys do, and the older girls would put makeup and what not on the younger boys. They were just wonderful children. I have no regrets about my children. I am very blessed. They blessed me with five beautiful grandchildren, and our grandchildren continue to increase the depth of love in this family. Love wins the battle.

What are your views on spirituality?

There is a God. All we have to do is open our eyes and look at nature, and we will realize that all this didn't just appear. God tells us we have no excuse for not knowing He exists, because of all the proof we have through nature. My views of spirituality hasn't changed since I was nine or ten and went to the alter as a little boy to accept Jesus Christ as my savior. I have total faith and very much desire to be a part of God's eternity. It's something that we have to work on every day of our lives . . . It takes every aspect of developing spirituality in one's life. It takes prayer, fasting, faith, and it just takes a continuous effort.

What do you think about teenagers today, and do you have any advice for them?

I love teenagers today; I don't have a problem with teenagers. The culture in which they live is the culture that they grew up in. It's no better or worse than the one I grew up in, just in a different time. The advice is to maintain character above everything else. Keep faith in God, trust God with every aspect of your life. Don't do anything without consulting God. I'm saying this because I didn't do it this way and I wish I had.

CHAPTER 15

Calling the Cows in Early

Biographical Information

Name: Paul Edison Weir
Place of Birth: Decatur, Arkansas
Date of Birth: December 7, 1918
Interviewed By: Tori Bonner, Brett Gathright, Dan Dillard, and Nenita Schell

Is Decatur located in the River Valley area?

No, it's up in the north part of the state.

How many siblings do you have?

Just one sister . . . She had a pretty foul mouth, and I don't know how well I dealt with that. They'd come up here on the weekends and keep us from going to church. She said, "I hope a steeple grows out of our house." [She] passed away six years ago. She married an Italian and had one little boy. My brother-in-law passed away, and then my nephew died . . . He drank himself to death. My great nephew drinks worse than his daddy did.

Does your family have any connection to "Weir Road" in Russellville?

Um, well, I don't know . . . I had a group of cousins that lived on that road a long time back. Center Valley is where my folks originally came from . . . Mother died when I was six days old, and Daddy died when I was five. My grandparents raised me. Actually, a lot of that stuff back there I don't know about. Daddy married again [when I was three]. He had another baby, a little girl. He died from appendicitis. It bursted before he knew anything was wrong and killed him. Of course, back then there was no penicillin or anything like that to take care of that . . . When he died, my aunt on my mother's side and my grandpa on my mother's side took me and my sister away from our stepmother. I didn't appreciate that. I didn't hear from my stepmother or my stepsister for 32 years. I still remember the day [the courts took me from my stepmother]. I kicked and screamed the whole time I was leaving the courtroom just trying to get back to her.

Do you think your mother's family just didn't care for your stepmom?

It was more jealousy . . . My aunt on my mother's side would have married my daddy, but . . . Anyways, after 32 years, a writing came up in the Little Rock and Fort Smith paper about me and my family. My stepmother, stepsister, and niece all saw it. I got a letter from my stepmother . . . I read it with my wife out loud. She asked me what I was going to do, and if I was going to write her back. I said, "No," because the first chance I got I was going to go and find her . . . My wife and I went to Fort Smith, because she had a sister up there that was getting married. This cousin took us to the house that my stepmother lived in. We parked across the street and went to the door. I walked up to the door and knocked a few times.

What happened next?

A lady came up to the door and I asked, "Is Francis Weir here?" She replied, "I am who you are looking for." I pulled out the letter and showed it to her. She said, "Yes, I am the one that wrote the letter, but you're not the Paul Weir that I knew." I told her that I could prove that I was the right Paul Weir by showing her the scars on me that she talked about in the letter. I have a scar on the back of my head. I had a rooster tail stuck up there, [which is strands of hair that naturally stick up on someone's head]. I took my daddy's straight razor and tried to get rid of it, and I got a little too deep. She got a phone call while I was there and told me that I'd have to come with her. It was her daughter, and the daughter asked if it was Paul Weir at the door. Her and her husband and her little boy came over, and we had a nice visit. Two weeks later, my sister and my two aunts were going to have lunch at my uncle's, so when we got home from church my stepmother was sitting in the driveway. She asked my entire family to come with her to eat . . . It was great reconnecting.

What is your perspective on going to church?

I will never forget, when I was 15, grandpa said, "You need to join the church." He never said a thing about being saved until then. I went to the Lutheran church and was confirmed with 12 others. Out of all the youngsters, I wonder how many of them got revived. I've thought of these others and wonder how many of them never mention to God, "Take my heart, and I need you in my life." I do . . . I go to a Presbyterian church now. I told my wife when she got saved, when we were dating, that if she wanted to join the church here, then I had nothing to quarrel about.

Religion was important to you early on, but was school?

[I went to school] in Mountain Springs. It was a little country school. I had to walk every morning . . . I had to quit school in the 7th grade to take care of grandpa. He had a bad sore leg. I guess it was diabetes. I didn't like school. I just had to go that's all. The one teacher we had made us spell every word before we said it. I'm a poor reader, but anyhow I made it. [Mountain springs only had] one building for 1st through 8th grade . . . I do think it's really important for younger people to have a higher education, so they can keep up with all these new technology changes and fast changing world that we live in now. Back in the old days, it wasn't important to have higher education because you didn't need it to work on the farm. You learned how to run the farm from your parents and grandparents back then.

Tell us about friends from when you were younger.

Do I have to? I had old friends and young friends. I had a lot of friends—a lot in school and a lot now. I have a friend to this day that I met in school that is six months older than me. For years, I never had any contact with him or anyone else. About five years ago, I knew some of his friends and family and they told him where I lived. We visited and went to eat lunch. We agreed that we need to get together more often.

Were any of the kids you were in school with bullies?

There was a lot of troublemakers when I was in school. I read about bullying in the paper awhile back. I never did feel like people should be doing that. There are so many good things you could be doing instead. I'll never forget, there was a little girl when I was younger, and she stole my jacket and threw it up on a pole. When I jumped up there to get it, I fell and ended up straddling her. Now that wasn't funny. There was another little girl that lived a mile or two up the road from where I lived. She'd walk to my house every morning, and I'd join her as we walked to school. We had a little trouble at school. We were having a play, and the teacher said that a little boy and my neighbor girl had to kiss in the play . . . The girl said, "No." The girl's daddy came to school to talk to the teacher about it. They ended up moving to a different school because of it. Needless to say, the teacher didn't work there much longer once her daddy got a hold of him. Anyway, we had one or two boys that were serious troublemakers. They were older.

What about entertainment when you were growing up?

The very first movie I saw was played at the church. We didn't have movie houses back then. The only way to get to the church for a movie was by horse, if you were lucky enough to have one. If you didn't have a horse, then you had to walk. Those were your only options . . . My grandfather bought me a horse when I was 16. I know it doesn't seem like much, but it was a great present. My first experience driving a car was when I was 18. I didn't buy my own car until I was 30. It was a half-ton Chevrolet truck . . . Anyway, we would get a group together and play movies all night. You know what was really fun? Community parties were so much fun. The girls would be up at 6:00 a.m., go pick cotton, go home early, and then get ready for the party. When the party ended at midnight, we would walk back home, and we knew that everyone was just going to turn around and start the day all over again at 6:00 the next morning. The girls working in the cotton field earned 75 cents a day. I know that doesn't sound like much money, but [that was] a lot of money then. Back then, if you wanted to buy a pair of shoes and some new pants, you had to work three days. I remember when we started getting dollar bills. Everyone thought they were rich.

Did any girls catch your eye at community parties?

I got my very first date when we had a community party. One of my neighbors insisted that I escort his daughter to the party. Her and her sisters were only allowed to go if I was with 'em at all times. I worked a lot on a farm, but I always knew if it was a planned party night because all of the girls would call the cows in early. I would hurry up with feeding the animals, milking the cows, and bringing up the firewood so I could go meet up with the girls. My grandmother never approved, but grandpa always gave in and let me take the girls to the party.

Did you ever fall in love with any of these girls?

No, no . . . They were all just friends. I had my eyes set on a girl that lived pretty close to me, but she was dating another boy. She ended up breaking up with him, because he would always get drunk and get himself into a lot of trouble. I stepped in after the break up and got her. We ended up staying together for a long time. We planned on getting married in a double wedding with my friend, Martin, and his wife. Me and Martin both ended up getting called into service, and it ruined our plans.

Did you serve in the military?

Well, Martin and I went to Texas for training and our exams. When I got my first physical exam, they tapped into my spine and fluid came out. I got detained for observation, and then got discharged from the military three months later. Martin stayed behind to finish training, and I came back to Arkansas. I married my sweetheart December 25, 1943, in the midst of World War II. Martin later came back on temporary leave, and his wife got pregnant. She didn't have any kind of prenatal check-up. Since Martin was still gone, I was the one that had to go run through the rain to try and find a doctor when she was ready to have the baby. The only doctor in the area refused to deliver her baby, because she didn't have any prenatal care or any kind of check—ups. I ended up talking the doctor into it. I explained to him that she had no way of getting a check—up, because her husband was currently in the service.

How was married life for you?

Well, we had two girls together. I remember when my first daughter was about to be born like it was yesterday. I got home from work about 6:00 that night. Right after I got home, my wife's water broke . . . I had to run half a mile to a friend's house to borrow their car to go and get the doctor, and my aunt. We made it back to the house just in time to deliver the baby. My second daughter was a different story though. I felt a lot more confident, since I already had a baby before. I volunteered to be in the room and cut the umbilical cord and everything that came along with the baby. I was wrong, I couldn't handle it . . . It was very miserable.

Do you have grandchildren?

Oh, yes. Between my two girls, they had 14 children total, so I ended up with 14 grandchildren and at least 24 great-grandchildren. I love them all so much. They're great kids. One of my great-grandchildren took me to their school for Grandparent's Day.

What did you do for work?

Farming . . . Me and three of my neighbor friends paid the same man to bale our hay when I was living on my farm. I was working then at General Mills in the poultry division [and] found out that someone was selling a used hay baler . . . I got together with the others, and we decided to buy it together. We had a disagreement over profit shares, [so we sold] the hay baler and I reinvested in my own equipment . . . I had to get a loan from my aunt to be able to afford a tractor to expand the business. I worked on the farm with grandpa when I was younger, and I was the President of the Farm Bureau and the Cattlemen's Association. I enjoyed both very much. I had a great time raising chickens and other animals on the farm.

Were home-raised animals always a big part of your diet?

Oh, yes. We rarely bought meat. We did it all ourselves. Any meat that we wanted, but didn't have, we got from a neighbor or from hunting. Grandpa gave me all of my advice for hunting. When I went deer hunting, the first thing he told me was to make sure and remove the sweat glands. He told me that if I didn't remove the sweat glands then I would ruin the meat, and no one would want it. I didn't hunt much for the meat though. I hunted for the animal hides. I made a lot of my money when I was younger catching raccoons and minks, and then selling their hides. I remember when all of the squirrels were migrating one year, when I was big into hunting. One squirrel about lost all of his nails trying to get away from me. My favorite story from those days is when I cornered five minks in a hole . . . Back then, selling two mink pelts got you more money than a full day's wages, so I loved to do it . . . We had a really big garden, and we raised hogs, chickens, and cattle. I had a 8' x 12' chicken house. I raised fryers and contracted with a local butcher for three pounds of fryers a week, but ended up getting him 25 pounds a week. Later on, people came straight to me for the fryers. Sometimes, we produced food. We stored corn, potatoes, beans, and asparagus in the root cellar.

Did you ever buy anything from the store?

The only things that we ever bought from the store was flour and sugar. Everything else we had came from the farm. We even made our own cornmeal. We had a poke salad plant, but I really never knew what it was for—some type of medicine I think. To avoid buying stuff from the store, we also kept seeds and traded them with our neighbors on a yearly basis. I've tried getting my daughter's into gardening, but they only see it as a hobby. They had small gardens where they just grew random plants. They say they don't have time for gardening since they both work outside of the home. To them, it is more time efficient if you just buy everything from the store, but when I was growing up gardening and raising animals was my livelihood.

Have you noticed any changes in wildlife in the last 50 years?

Yes. I believe the wildlife is getting thicker. They come down in the valley now. They pass through my property every single morning this time of year. Me and my daughter think they're just hungry and are looking for food. We put out food for all of the wildlife all the time. I love seeing the animals around, since I don't hunt or raise animals for meat anymore. I do love eating young deer, because it has good meat. Buck is tough. It's not very good to eat.

Attitudes on race have changed a lot recently. Is that a good or a bad thing?

The way I see it, we all have a heart. We all have the need to live, and best of all, we were all created by God and therefore are equal.

Overall, what has been the best thing about your life?

Well, besides marrying a wonderful woman and having so many kids, grandkids, and great-grandkids, there are some other things. Like I said earlier, I was the President of the Farm Bureau and the Cattlemen's Association, and I regularly went to church. I also started a local fire department . . . I loved all of my life. All parts of people's lives have a reason for being that way. Everything was nice. I don't have anything to grumble about. I have had a long, happy life full of great friends and family. As long as you treat everyone the way you want to be treated, then you should have a happy life . . .

As you have gotten older, how have your relationships changed?

I play with my great-grandchildren to keep my mind sharp. I have a lot of friends that are in nursing homes that I go and visit quite frequently. It is sort of strange seeing them in there though, because I am older than most of them. They always seem happy that I stop by though . . . I would accept [living in a nursing home]. I think that if I lived [in one], I would be able to do a pretty good job at cheering [other people] up. A lot of people aren't very happy about having to rely on other people, but I don't think it'd be the end of the world . . . I've outlived a lot of my friends and family. I've lost a lot of really important people in my life, but I know they've gone to a better place. I am very proud of my age and that God has given me a chance to live this long. I try to look at life in a positive way. Everything is beautiful, and everything happens for a reason. I am ready to go anytime and join my family, but in the meantime, I am going to keep living life to the fullest.

REFERENCES

Blevins, Brooks. 2009. *Arkansas/Arkansaw: How Bear Hunters, Hillbillies, & Good Ol' Boys Defined a State*. Fayetteville: The University of Arkansas Press.

Quadagno, Jill. 2014. *Aging and the Life Course*. New York: McGraw-Hill.

Ulsperger, Jason S. 2001. "Perceptions of the Aged in a Young Adult Population: The Importance of Exchange in Intergenerational Relationships." *Southwest Journal on Aging* 17(1/2):31-40.

Ulsperger, Jason S. and J. David Knottnerus. 2011. *Elder Care Catastrophe: Rituals of Abuse in Nursing Homes—and What You Can Do About It*. Boulder, CO: Paradigm Publishing.

Ulsperger, Jason S., Cole Smith, and Kristen Kloss Ulsperger. 2009. *River Valley Reflections: Heritage from the Halls of Long-term Care*. Philadelphia, PA: Xlibris Publishing.

Ulsperger, Jason S. and Kristen Kloss Ulsperger. 2008. *Voices of Pope County: Gerontological Perceptions of the Past and Present*. Philadelphia, PA: Xlibris Publishing.

Ulsperger, Jason S. and Kristen Kloss Ulsperger. 2011. "Working with the Words of Elders: A Guide for Teachers and Librarians Developing Oral History Projects." Arkansas Libraries 68 (3): 4-7.